Tarantulas & Marmosets

An Amazon Diary

Tarantulas & Marmosets

An Amazon Diary

NICK GORDON

metro

First published in Great Britain in hardback in 1997
by Metro Books (an imprint of Metro Publishing Limited),
19 Gerrard Street, London W1V 7LA

This paperback edition published 1998

British Library Cataloguing in Publication Data. A CIP record of this book is available on request from the British Library.

ISBN 1 900512 37 8

10 9 8 7 6 5 4 3 2 1

All photographs taken by Nick Gordon with the exception of production shots of the Survival team, which were taken by Gordon Buchanan and Stephen Terry.

Maps by Michael Hill.

Typeset by Sally Brock, High Wycombe, Buckinghamshire.

Printed in Great Britain by CPD Group, Wales

To my two girls,
EMMA AND CHARLIE

Preface

This book condenses more than a decade of making wildlife films into a few hundred pages. I would not have experienced any of it without the tireless back-up and support from so many other people whose names would probably fill another volume. I am especially indebted to Survival Anglia and their entire team, with whom I have worked over many years, for their dedication, professionalism, continuing faith in my work and the friendships that have grown from our relationship. I shall also be forever in debt to Nick Peake who believed in me, under difficult circumstances, when we sold my first film to Survival Anglia and thus first opened the door for me to the extraordinary world of making wildlife films.

In all, I ended up spending a total of 11 years in the rainforests of South America, making six films for Survival's wildlife series – Survival being the main providers of wildlife films for Independent Television. The films are: *Giant Otter,* made on my first visit to South America in 1985 (producer: Caroline Brett; film editor: Michael Holmes); *Tarantula!* (producer: Caroline Brett; film editor: Rob Harrington); *Creatures of the Magic Water* (film editor: Mark Anderson); *Web of the Spider Monkey* (film editor: Howard Marshall); *Gremlins – Faces in the Forest* (film editor: Mark Anderson); and *Jaguar – Eater of Souls* (in production).

Acknowledgements

Being away from home for such long periods can, and at times has been, hard on the spirit, and my family and friends have played a more important role in *my* survival than they could possibly imagine. To all of them, my deepest thanks and love. I hope that this book explains my long absences and, perhaps, partly justifies them.

I owe a special debt of gratitude to Sir David Attenborough for graciously agreeing to support my first book with his quote. To Caroline Brett, Kip Halsey and Alison Aitken, my appreciation for their constant support and encouragement, and to Rick West for introducing me to the giant tarantula – and much else! Thanks, too, to my long-suffering assistant Gordon Buchanan, whose dedication and humour were a constant support. I also thank Marc van Roosmalen and his wife, Betty, for their kind hospitality over the past three years in Amazonia. But for their involvement the last three Brazilian films would not have been made. Finally, thanks to Nigel Blundell, who set off my writing and to Susanne McDadd and the team at Metro for nursing me through it all, mostly by way of a satellite communication system that more often than not carried only the calls of the monkeys in the background, my voice just a digital blur.

'Here at last is the book that so many who have enjoyed natural history television programmes have been waiting for – the story of what goes on, not in front of the camera but behind it, told by one of the master practitioners.'

Sir David Attenborough

Contents

Chapter One

Flight into the Unknown

25 September, 1991

The bush plane bobbed gently along the grass strip in the cool, early morning tropical air. Frayed ropes attached me to the plane's opening where two side doors had been removed to allow me to film the spectacle of the sun rising over the Amazonas rainforest in Venezuela, South America. I felt uneasy as I looked at Junior, my pilot, and wished he was twice his 25 years in age and experience.

Junior gave the single engine full revs and we hurtled along the runway, the noise of windrush shattering the peace.

Within 15 minutes I was looking at an awe-inspiring scene. Dark green forest stretched out below me as far as the eye could see. The glow of the unborn sun was spreading from the eastern horizon, with fingers of blazing light coming towards us as though they were going to snatch us from the sky. Clouds of vapour enveloped the forest canopy 500 feet below, allowing just a few tree crowns to poke through in a ghostly vision. The gloom of the silhouetted highland mountains added to a scene so special that my mouth gaped in wonder. Soon the mists would begin to lift and disappear and we would see the forest in all its splendour.

I was in southern Venezuela, flying towards a magnificent pinnacle of rock rising almost vertically some 4,000 feet from the jungle below. This was the sacred mountain which the Piaroa Indians who live in the area call Wahari Kuawai. The forest shrouding its steep slopes and leading up to the Rio Sipapo, a tributary of the Orinoco river, is home to the Piaroa Indians, one of the few groups of indigenous peoples that still inhabit this part of Venezuela. Piaroa legend tells how the mountain, Wahari Kuawai, was once a tree full of all the fruits of the forest. A boy called Tunga was sent by a devil to cut the tree down to provide food for the men of the jungle. After he had felled the tree a tapir came along to eat the fallen fruit. Tunga chased the tapir away and turned it into stone. According to legend this now forms the smaller mountain of Wichuj to the south of Wahari Kuawai.

The southern region of Venezuela forms the northern extremity of Amazonia, the greatest forest on earth. Its wilderness covers more than 2 million square miles of land and it spans eight countries. This area is remarkable for its many *tepuys,* spectacular isolated sandstone mountains found throughout this part of the world.

In 1884, the explorers Everard Im Thurn and Harry I Perkins led a Royal Geographic Society expedition to Roraima, the most famous of these *tepuys.* On 18 December they reached a point close to the summit and Im Thurn wrote in his journal: 'We have now reached a spot where one long climb will take us to the level of the summit and we shall behold that which has never been observed since the beginning of the world.'

A Royal Geographic Society lecture about this expedition inspired Sir Arthur Conan Doyle to write one of the best science fiction novels ever written, *The Lost World* (1912). He told of a mysterious mountain whose vertical walls rose from jungle and at whose summit was a lair of dinosaurs isolated from an unsuspecting world.

Just over 100 years later we were circling our own lost world, the *tepuy* of Wahari Kuawai, my mind swimming with images from the tale that had so enthralled me as a child. In these incredible surroundings it was easy to blur the distinctions between fiction and fact and I conjured up pictures of creatures existing undiscovered on this extraordinary mountain.

Junior and I were among the few outsiders to come so close. As our little aircraft flew around the summit, dwarfed by the size of the mountain, I was reminiscing in my dream world. Suddenly the sun appeared on the horizon with a second's burst of sapphire light, silhouetting Wahari Kuawai before it. I was jolted back to reality. I began to film the dawn of a new day, deeply aware of the privilege of being the only person in the world – apart from Junior, of course – to be witnessing this magical scene. He gave me the thumbs-up sign, a broad grin giving away his own delight.

We were circling about 500 feet below the summit of Wahari Kuawai when

Junior shouted something about a cave, turning the aircraft so steeply that my camera box began to slide towards the opening. I lurched forward to pull it back and was horrified to see us heading straight for the rock face. Within seconds there was nothing in view but dark jagged rock. I clung to the door frame wide-eyed with fear. Seconds later Junior threw the little plane hard over again, and we flew along the rock face – now only 150 feet to our right. I was trying simultaneously to hold on to my camera and keep myself from falling out, enraged by his dangerous manoeuvres. There was enough slack in the old ropes around my waist to mean I was in danger of sliding.

Junior pointed at something and I saw a tunnel. It was only a fleeting glimpse, but I could see right through the mountain. Fascinating though this was, I'd had enough excitement for one morning and was only interested in moving safely on.

Living in the darkness below the forest roof of the Piaroa Indians' emerald world is a creature that could have come straight out of Conan Doyle's tale. It is the largest tarantula on earth, known to science as *Theraphosa blondi.* This spider, and the Piaroa's relationship with it, was to be the subject of my latest film. The expedition would take me deep into the heart of the Amazonian rainforest to live and work among the Piaroa and Yanomamo Indians of Venezuela. As one project so often leads to another, I was also to cross the border into Brazil, set up home in the forest, and film the giant turtles, marmosets and other wildlife of the River Amazon and its tributaries.

Chapter Two

Behind the Scenes

It was April 1987 and I had just returned to the United Kingdom after two years in Guyana, Venezuela's eastern neighbour. I had been living in a mud hut on the edge of the Rupununi Savannah, Macusi Indian territory, filming my first Survival Special, *Giant Otter,* about the largest otter in the world. I had lived, breathed and dreamed otters for so long that I was now ready for a fresh challenge.

My heart was captivated by the South American rainforests and I was keen to return, but the Survival team had other plans. They wanted me to go to the Gola rainforest in Sierra Leone in West Africa, and shoot another Survival Special about wild chimpanzees, *Tiwai Island of the Apes.* With the project in West Africa scheduled to take a year and a half I did not know when I would be in a position to return to South America. Unknown to me fate was already taking a hand.

During my 16-week break in the UK, a couple of books were published containing photographs of mine taken in the Guyanan rainforests. This led to Rick West, a scientist from British Columbia, contacting me for any information I had about tarantulas in Guyana. I sent him some of my pictures of spiders and whilst I was in West Africa we kept in regular contact.

Rick West is a world authority on tarantulas. From our correspondence, and the few times we managed to speak by telephone, I could tell that he was a kindred spirit, enthused by his work with spiders and the forest environment. During one of our exchanges Rick asked if I might be interested in making a film about the largest spider in the world, the tarantula known as *Theraphosa blondi* which lives only in the rainforests of South America. He mentioned in passing that a group of Venezuelan Indians ate this mighty arachnid.

He told me that they worshipped it and regarded it as a delicacy and that there was some kind of ritual they performed before hunting it. I was enthralled.

As I make a living from filming wildlife, many ideas for films rattle around my office at home on the island of Mull in Scotland. Inevitably, the majority never materialise. But as soon as Rick mentioned that the Indians ate the biggest tarantula in the world, I knew there just had to be a good film to be made. It was such a bizarre tale. The best wildlife films always revolve around a strong storyline and this was one of the best I had ever heard.

During the summer of 1991, the African adventure behind me, further research reinforced my instinct. Specialists at the Natural History Museum in London were just as fascinated by the story as I was and a call to the British Tarantula Society revealed that it, too, was excited by the prospect of a high-profile film being made about the species that captivates its members. Everyone I spoke to about the idea – except my mum – was delighted that I was going to take on the project.

There were some problems, as there are with any film. I was skilled in long lens work – sometimes filming 150 or more feet away from the subject – but had never filmed extremely close up. Although I was keen to have a go, it is specialist work. Like a poster artist turning his or her hand to miniaturism, it's a whole new discipline. (*Tarantula!* eventually won five international awards, one of them for the photography, so in the end I felt I had really achieved something.)

Another problem was the fact that this tarantula had never been filmed or even studied before, so that it was extremely difficult to write a film treatment for Survival. Treatments, or outlines, are film sales pitches, and have to be both good and credible to attract the financial backing of a television production company. Rick assumed that for the purpose of writing the treatment most of the giant tarantula's behaviour patterns would be similar to those of other better-known tarantulas. Nevertheless, I was excited by the possibility of discovering and recording new behaviour for the first time. Little did either of us realise during those long-distance chats between Canada and Scotland just how fascinating and unique the sequences that we would eventually capture on film would be.

And another thing I had not reckoned with was the political factor. It was 1991 and Margaret Thatcher and her Conservative government were in the middle of shaking up the British television industry. All the franchises to broadcast had been put up for auction to the highest bidder and none of the major Independent Television companies knew whether they would retain their licences. Rumour was rife within Anglia Television, Survival's parent company, that it might lose the franchise to another bidder. Because of this the management had decided not to commission any new films from its teams, including me, until the bids had been resolved.

I wasn't going to give up easily, and decided to see Survival's executive producer, Graham Creelman, personally. Our meeting went well, despite the uncertain climate, and he agreed with me that the giant tarantula story line was so strong it should be commissioned. I left his office so relieved and excited I whistled all the way back to my car.

We went to work immediately and soon heard from our contact man in Caracas, the capital of Venezuela, that visas and work permits would be forthcoming. In my usual state of excitement at the start of these expeditions I always forget to remind myself that with wildlife filming things never go smoothly. A million and one things had to be hurriedly organised.

My filming equipment was being serviced and repaired following my recent West African adventure. The damp from the rainforest in Sierra Leone had wreaked havoc with my cameras and lenses and all nine lenses needed new optics because of fungal growth. I hoped futilely that Amazonas would be less harsh on my equipment. Many other items had to be procured – especially 20 pounds of silica gel – to try and at least retard the effects of humidity. As we were going to be so far from civilisation, there was a mountain of equipment to be packed, from my solar battery charging system for the cameras to medicines, ointments, pills, anti-fungal creams, antiseptic powders... Survival producer Caroline Brett even sent me the latest snake bite treatment kit! I hoped that none of these items would be needed.

No amount of expensive equipment could make up for the cameraman's most essential aid – an assistant, especially one of Gordon Buchanan's calibre. Gordon started his career with me in 1989 when he was just 17 years old. Survival receives many letters from people wanting to work as assistants and I had a list of 100 hopefuls from every corner of the British Isles. As it turned out the successful applicant came from much closer to home.

I remember the occasion well. It was one of those crisp spring days and I was sitting on my balcony breathing the clean air and enjoying the views of the Western Isles. Suddenly there was this young lad who lives less than a mile away from my house, winking at me. Confident or what? He had brought his biology teacher with him to talk about the pros and cons of leaving school to take on the job of being my camera assistant. Only two weeks earlier he had simply knocked at my front door and asked if there was a job going.

Looking at him sitting there so innocently with his whole life ahead of him, I thought back to when I was leaving school. I remembered that although I had good A-levels, I opted, against everyone's advice, not to go to university and decided instead to get a job. The reason was simple – I had just met a girl with whom I had fallen in love. It wasn't that going out with her was incompatible with going to university, but having been incarcerated in an all-boys' school for five years, my first taste of freedom and the opposite sex was too much to give up. Reality dawned three months later when I saw her walking down Blackpool promenade arm-in-arm with another boy. I was gainfully employed earning £4 a week as an office boy in a firm of chartered surveyors in Blackpool. Making tea for the partners and collecting rent from an abattoir was proving less exciting than I had imagined.

With the benefit of hindsight I was far from convinced that leaving school was the best thing for Gordon. I told him to imagine his worst nightmare and then double it. I also told him to consider what he would do if he didn't like it and of the sort of jobs he could get without qualifications. But if he felt he still wanted to give it a go, he'd got the job.

Nothing I could say would dissuade Gordon, and he joined the team. The first stop for us was the Sierra Leone trip, at that time the poorest country on earth. We spent our 18-month expedition living on a disease-ridden uninhabited island in the middle of nowhere. Not only was Gordon misguided, I thought, but I must be completely bonkers, too. Forty-eight hours after leaving Scotland the 17-year-old had caught dysentery, was vomiting every hour and had seen his first dead body. Shocked doesn't do justice to what he must have felt. There were only the two of us, and from the moment we arrived it was a seven-day-a-week job – no time off in the normal sense. I was well used to this kind of dedication, but Sierra Leone unsettled me, too.

The following year we had to flee Tiwai, the tiny forested island where we were filming chimpanzees, as a rebel army from Liberia invaded the area, killing 300 people. According to some missionary friends, fresh from the horrors of seeing human heads impaled on wooden stakes, we had missed that fate by some half an hour.

Gordon's homesickness didn't really leave him until the second half of the African project, some nine months later. We had returned to the UK to go through the film, and to spend a few weeks with our families and friends. Gordon discovered two things that I believe shaped his future. Almost everyone he met said they would have given their right arm for his job, and most of his closest friends had moved off to college and were making their own way in the world, like him. There is no doubt that their envy, in the nicest way, was like a beacon for Gordon and he realised what a wonderful opportunity he had been given.

My fears about taking on an assistant with no experience faded fast. And – from a purely selfish point of view – he learnt to do things the way I wanted them done! This strappingly-built lad had a sense of humour to match his physique and we were to become close friends. I came to rely upon him in a way that took a great burden off my shoulder from building 150-foot scaffolding towers into the forest canopy to negotiating with people on my behalf, whether native workers or government ministers.

Five and a half years and three films later, Gordon left to make his own films. He has every right to look back on our time together with pride and he deserves a medal for putting up with me for so long.

Sometimes as I watched Gordon work, I wondered how this young man had made it through the gruelling experiences we encounter in our business. Living for very long periods in tents, with no other person for company, can be a strain, and we had moments when we probably felt like killing each other. But Gordon was very good at confronting a problem festering between us. 'C'mon, what's the matter, dad?' he would say, shoving a hot mug of Horlicks through the mosquito netting door to my tent and this would usually break the ice. My only long-standing grievance was that when we walked through isolated villages or Third-World cities, Gordon attracted all the girls while I attracted the beggars. It's an unfair world.

On a personal note, the Venezuelan tarantula project was to be something of a life-saver for me. Many people imagine that making wildlife films is exceptionally glamorous – free travel, all expenses paid and so on. But the work has its unglamorous side, not just photographically, but also in maintaining a life outside filming.

In April 1991, I returned from Sierra Leone to find my other world falling apart. My wife, Ann, told me she was leaving me to set up home with someone else. Ann and I had been together for 15 years and had a beautiful four-year-old daughter, Emma. Anyone, and I know there are many, who has experienced such trauma will be able to identify with the pain and desolation that one feels, particularly when the break-up is totally unexpected. Ultimately, like so many failed relationships, neither of us had talked about feelings. And, of course, my long absences for work didn't help. I had never felt spiritually lower.

It took strength I never knew I possessed to motivate myself to prepare the tarantula project when all I wanted was to make things right again and for us to be a family once more. I was consumed with fear and, for the first time, lacked confidence and felt lonely. I knew only too well from past projects how thoughts of home and family, the exchange of letters, and Ann and Emma's visits to join me in the field always gave me great inner strength. It was the first time I had ever asked myself if what I was doing was worthwhile.

Preoccupied with these concerns, the summer of 1991 in west Scotland disappeared all too quickly. I was still stamping on the African cockroaches which crawled out of my equipment cases when the four months of research, preparation and late-night transatlantic telephone calls culminated in my departure for Amazonia. I had never made a wildlife film without first doing a reconnaissance of the country and making sure that the substantial financial investment that these films involve would be secure. In this case there was no time and I had no choice. I had to go with it on a gut feeling.

In September 1991, Peter Schofield, head of Survival's production unit, met me in Norwich to assemble all the final bits and pieces as well as the boxes of precious film that from here on had to be handled and cared for like babes in arms. Tina, Peter's wife, provided the delicious last supper on the eve of our flight, an event that had become a ritual for me and Gordon before we set out on our far-flung expeditions. We all sat round the dining table savouring every mouthful of home-made steak and kidney pudding, followed by apple crumble and custard. Tina's home cooking is a source of nostalgic comfort for us when we are in our tents or huts deep in the forest, eating our Spartan diet of rice or leaf or even worse… Tina always gets straight to the point. 'What on earth drives you two? Why go and live in mud huts surrounded by snakes?' she would ask. But while eveyone pulls faces at the very mention of such things, they all know that it's in our blood. They also know that despite this we do miss our home comforts. As I climbed into bed I thought about our splendid send-off and our last such feast – not to mention how this would be my last night under a duvet – for a year or more.

The next day, I faced the last-minute headaches of getting 1,000 pounds of filming equipment through Heathrow airport. Then, as our Boeing 747 climbed through the skies bound for South America, I prayed that I would be able to cope. I was relieved to be getting away and to have something on which to focus my mind, but there was also an inner fear that I was running away. And what about my little girl? I had no answer to that, just a big lump in my throat whenever I thought of her.

During the long night flight I inwardly debated these confusing thoughts over a couple of drinks. I reached the conclusion that as I had to work I might as well work at something I love doing, making wildlife films.

Take One: Venezuela

Chapter Three

Our Man in Caracas

The descent into Caracas presented a magnificent aerial view of the city. Caracas is situated along a narrow valley flanked by mountains and we could see their summits collared with white puffs of cloud. The view even distracted Gordon from chatting up the girl on his left. As I looked down I thought of the early explorers who had come this way, like Columbus in 1498, who discovered the river mouth of the Orinoco, the fifth longest watercourse in the world. How different it would all seem to them today with the miles of concrete jungle and the millions of people clustered like ants along this stretch of coastline.

Our jumbo jet touched down during the late and stifling hot afternoon. We were met by 'our man in Caracas', a charming Venezuelan, the well-known explorer and naturalist, Charles Brewer-Carias, who had escorted Prince Charles into the Venezuelan rainforest the year before. (Television productions usually have a local liaison person when working abroad, as their contacts can help to short-cut much of the inevitable bureaucracy.)

Dr Brewer-Carias bore a close resemblance to James Coburn in the film *The Magnificent Seven.* Aged about 50, disgustingly tall, with looks many would kill for, he was also a thoroughly decent bloke. He had organised for us to be whisked through immigration and customs as VIPs but one thing even the indomitable Charlie could not save us from were the vulturine baggage handlers. They had spotted our 38 silver-coloured equipment boxes and were moving around us in ever-decreasing circles. Once one of them had landed, the normal Third World free-for-all ensued. Two of the many porters conducted a tug of war over my camera box. A pile of eight smaller cases crashed off one of the trolleys onto the floor just as an official arrived to sort out the men. It is never any different, yet I always hope it will be.

After an hour and a half driving through horrendous traffic from the airport to the city centre, some 11 miles away, breathing in copious quantities of carbon monoxide fumes, we settled into our hotel rooms. Charlie left Gordon and me feeling a bit fazed by it all, and we allowed our jet lag to put us to sleep for the next ten hours. The last thing I saw before

I fell asleep was a cockroach, the size of a boiled egg, scuttling up the wall next to my head. And the last thing I remember thinking was that 'Welcome to Venezuela' should have been painted on its back.

The following morning we were due for breakfast at Charlie's house. Although it was only two miles away, it took us an hour by car to get there, crawling bumper to bumper through bustling streets.

Charlie's house was in the suburbs and surrounded by a high security wall with a large metal gate. He met us in his front garden where a handsome tame white eagle was perched on a small tree with bare branches. I chatted with Charlie about the project over tumblers of freshly squeezed passion fruit juice – one of the simple pleasures in these tropical climes. Later he introduced us to his wife, Fanny, a strikingly beautiful young Venezuelan girl. Charlie's study is an organised jumble of library books, artefacts from his many expeditions, countless preserved specimens of giant beetles and scorpions that would make most entomologists drool – and lots more. I felt immediately at home. I scanned the hundreds of books on show. One section was labelled Jungle Survival Techniques – the first volume I pulled out was titled *How to Deliver the Fatal Blow*. Its pages were filled with graphic illustrations of how to kill another human being. Could be useful if a producer causes problems, I thought. Back to business.

Several glass vivariums were filled with poisonous snakes and what looked like extraterrestrial beetles. In one of them I had my first sight of the subject I would be filming over the next 12 months, *Theraphosa blondi*, the giant tarantula. Charlie had kept this live specimen for a few months. It was a female and, when I asked him how you sexed a giant tarantula, he explained that the female's abdomen is much bigger than a male's – in this case about the size of a ping-pong ball. Another difference is that the males have a kind of spur or hook on the inside of their two front legs.

The tarantula was truly something to see. With a leg span of 10 inches, and a body weight of 4 ounces, she was a dark chocolate brown colour and surprisingly furry. My mum – indeed most people – would have had a seizure if this creature had crawled out from under her sink.

As Charlie lifted the vivarium's cover, I heard a hissing sound similar to the noise made by some snakes. Charlie pulled my head away from the opening. This breed of tarantula has a highly developed defence mechanism consisting of special hairs that cover the spider's abdomen. When it feels threatened, it rubs its hind legs, which have stout spines on their inner surface, across the sides of its abdomen. This friction sends a cloud of microscopic barbed hairs into the air above it. These urticating hairs, as they are called, are extremely irritating to humans and cause terrible discomfort. They can become lodged in the surface of our skin and make it itch badly. If certain animals inhale these hairs their respiratory tracts become so inflamed they choke to death.

Once the air had settled I moved in for a closer look. So this was 'incey wincey'. We were going to have to catch and keep captive some specimens from the forest to allow me to film the special close-ups of its mating behaviour. I saw her three-quarter-inch fangs and silently thanked Rick West for taking on the job of being my scientific adviser for this film. He would divulge the secrets of handling and keeping alive our specimens before leaving us to it.

Rick arrived the following day and I met him at Caracas airport. It's a strange feeling only knowing someone by telephone and letters for such a long time and then meeting them face to face. I had built up an image of Rick over the last two years, as tall, dark, spindly and thin-faced. But he was nothing like I had pictured him. He was freckle-faced, ginger-haired and double the weight I had imagined. He plodded towards me with a tired smile on his face, weighed down by bags of bug-collecting gear. 'Hi guys, I'm knackered. How yer doin'?' were his first words.

After telling each other half a dozen times how great it was to meet in the flesh, we took a taxi to the hotel, babbling away about the project. 'Jeeez, I never thought we'd make it, guy, especially without permission.'

'What the hell are you saying, Rick? We've come into a foreign country *without permission*,' I shrieked.

'Only joking, guy, Charles has sorted everything out.'

As I was to find out, Rick's usually monotone voice belies his droll sense of humour. We hit it off immediately, though. My paranoia was slowly rising as I wondered about another little matter, that of making the film. He unpacked and showed me his spider-collecting gear – we must have looked like schoolboys comparing stamp collections. Later that day Rick met Charlie at his home. After two days with Rick, my file on the natural history of this incredible spider had filled out considerably. We now knew that *blondi* was a 'fossorial' spider. Rick matter-of-factly explained that it will dig its own burrow, although it would also quite readily take over existing forest floor dens excavated by small rodents or mammals. We had also learnt that the spider's strategy for capturing prey was to sit and wait outside the entrance to its burrow from sundown to sunrise, and pounce on any unsuspecting creature that passed too close. It would generally be found near creeks, swamps or rivers. The male spider could be found wandering the forest floor during the day or night, a tunnel-visioned sex machine, his sole purpose in life to find a receptive female and mate with her.

'Wow, what a simple but meaningful life,' Gordon put in. Perhaps, I commented, Gordon should be re-classified scientifically as a member of the arachnid family.

Both the female tarantula and the burrow are detected by means of specialised chemoreceptive hairs on the male's legs. Gordon asked Rick if he thought in his professional scientific opinion that this might mean that girls with hairy legs might... Amid the laughter I was happy knowing we were all going to get on just fine.

After mating with a male, the female giant tarantula produces an egg sac the size of a tennis ball and jealously guards it in her underground lair. From birth, males and females mature in two to three years but the males die shortly after maturity whereas the females can live for several more years. 'Piece of cake, let's go and film it,' I said to Gordon and Rick, not believing for one minute that any of it was going to be quite so straightforward.

We still had two more days in the city making last-minute purchases. Caracas was stifling, although I had been told by many people that it was often cooled by the sea breezes coming down the valley. No such luck for us. The valley in which Caracas stands is narrow and long, ten miles long in fact, and crammed to bursting point with ten million people and what seemed like as many cars. I couldn't compare it to any other city I had ever seen, the bustle, heat and smells made it unpleasantly unique. Imagine the contrast to present-day Caracas at the time when the Spaniard, Diego de Losada, invaded in 1567. At that time only a few Toromaima Indians were settled here, and they had named it after a wild grass that grew on the edges of the valley's rivers, Caracas. Today, it is an ultra-modern, high-tech, high-rise city centre surrounded by slums sprawling perilously up hills and mountain sides. When bad storms or landslides hit the area, the shanties collapse, sliding away in a torrent of mud or rubble. I found the dichotomy of wealth and poverty in Caracas shocking – the acute poverty and squalor a recipe for social tension.

The military were more evident in the streets than seemed proper. They had good reason to be, as events turned out not so long afterwards. Gordon and I had fled a civil war in Sierra Leone and I did wonder whether we had jumped out of the frying pan and into the fire by coming to Venezuela. Chatting to Charlie about the city's make-up made us all feel more uneasy, as he had been a government minister and knew what was going on behind the scenes. He told me that for him the city had lost its soul and he was worried that something was about to happen. I believed him. I wanted us to get out of the place as quickly as possible.

Our base camp had been organised in the capital of Amazonas, the port and town of Puerto Ayacucho, some 300 miles south of Caracas on the edge of the Orinoco river. It was also within two days' travel of the area where the tribe of Piaroa Indians who still lived in the rainforest were clinging on to their way of life.

The day we left Caracas for the Federal Territory of Amazonas, Charlie took us to his private colonial country club for a farewell lunch. What food! Rick, Gordon and I didn't need any encouragement for we all knew that our diet was to undergo a drastic change once we left Caracas. The club was only available to the rich and influential and, looking around the sumptuously decorated reception rooms, I imagined a scenario where its members were holed up inside, armed to the teeth, keeping the insurgent hordes at bay.

Sitting with Charlie in the restaurant, sipping ice-cold fruit juices, a strange memory came flooding into the present. During the many months I had been inside my film hide on Tiwai Island in West Africa, waiting for a glimpse of the wild chimpanzees that live there, I filled the empty hours reading. A friend in England had given me a paperback recommending it with the words, 'You'll enjoy this, it's by another nutter who likes doing the things that you do.' I didn't appreciate the analogy, but it was an entertaining read.

The book, *In Trouble Again*, was by Redmond O'Hanlon. It was an account of his journey in Amazonas, Venezuela, to find the Yanomamo

Indians. I remembered that the author mentioned an explorer in Venezuela and it suddenly struck me that he was referring to Charlie. At the time of reading *In Trouble Again* little did I realise that only one year later I would be having lunch with the explorer himself and following in part the same route that O'Hanlon described.

The flight from Caracas south to the town of Puerto Ayacucho was fascinating, particularly the view of the Orinoco. At Ayacucho, the river divides Venezuela from Colombia. On the Colombian side there was clear savannah land as far as one could see and yet immediately the waters touched Venezuelan territory there was a vast expanse of forest that I knew stretched east towards Guyana and south into Brazil.

Our tatty, national-fleet aircraft hit the airstrip so hard that the cabin filled with the passengers' screams. We bounced twice and I cursed the unseen pilot at the controls while many passengers booed loudly. The heat at Ayacucho was even worse than at Caracas. In the stifling atmosphere of the airport, where the air-conditioning had broken down, my body seemed to be dripping away at about a pint a minute. It was extremely uncomfortable standing around in clothes soaked with sweat. As I was trying to count the equipment cases, a hand slapped me on the back and what sounded like a throat lined with sandpaper said '*Ola*'.

Julio could have stepped straight out of the Butch Cassidy film set. He was about 35, short, dark, with a droopy Mexican moustache, a happy-go-lucky face and a pot belly. He could also – thank heavens – speak excellent English. He was the sort of chap you immediately felt at ease with, and I took an instant liking to him. He had organised some helpers to get the mountain of film equipment to a waiting truck. As we humped the boxes out of the airport and into the open, the full force of the heat hit us.

'Haay, like an ooven,' Julio said slowly. 'We'll stoop on the way to the camp for a dreeenk. Haay, do yoou play poowel?' he asked. So Ayacucho had some recreational facilities. In fact, Julio always spoke very slowly, as though he considered every word first. But he, like Rick, was nobody's fool. 'Poowel' was the passion in his life, apart from his family and Venezuelan beer, in that order it seemed.

The Venezuelan territory of Amazonas covers about one fifth of the whole country and has vast frontier areas with Brazil, Guyana and Colombia. I found it amazing that Amazonas, with 30,000 people, contains less than two per cent of the country's population. More than half of these 30,000 are Indians. They come from 20 or so tribes, each with their own language and culture.

Around Ayacucho the Orinoco is characterised by large and dangerous rapids. From Ayacucho's port you cannot go up or down river, only overland to points beyond the rapids before you are able to travel again by boat. I would be getting a much closer look at those rapids, but for now I was happily unaware of that.

Anyone can be forgiven for wondering why Puerto Ayacucho was ever founded, being a thoroughly unremarkable development in one of the most remarkable areas on earth. The cost of living is extremely high, because almost all commodities have to be freighted in by air. 'So why was it built, Julio?'

'Haay, it's a yooung place,' Julio told me. Ayacucho was originally built in 1924 as an engineers' camp, when the government decided to link the two extremities of the Orinoco rapids by road. Prior to 1924 it just consisted of a river, rapids and rainforest. More poignantly, that date, I knew, heralded the worst period for the indigenous peoples who were now suddenly within reach of the 'civilisers'.

Today, Puerto Ayacucho is like any other Amazonian stopping-off point, whether in Venezuela or Brazil. It has become a Mecca for adventure travellers. Eco-tourists, backpackers, drifters and various shady-looking characters all conspire to give such places a similar look and feel. In Ayacucho many of the locals have seen commercial opportunities with the advent of green holidaymakers – and my goodness are some of them green. Number one travel rule in any foreign country is always to look purposeful even if you are lost!

Our base for part of our expedition was a camp on the edge of Ayacucho built specifically for people wanting to get a closer look at the great forest – those green holidaymakers in fact. Campamento Genesis is owned and managed by Jorge (pronounced 'horhey') Contreras Guerra and his wife Adrienne. The grounds around the camp are well kept and we would often hear a yell and see one of the gardeners killing a poisonous snake with a machete. The camp accommodation itself is built out of mud bricks, billet style, forming a three-sided courtyard with about 14 rooms, each with a small shower room. Although spartan, our room had comfortable ex-army barracks beds, and an air-conditioning unit that sometimes worked. There is a communal dining room and kitchen where the maids appear each morning with ham and cheese platters – food we would come to know well. On one wall facing the diners is a huge picture of the Piaroa's sacred mountain – Wahari Kuawai.

Jorge was a 39-year-old Panamanian who had travelled the world as a trouble-shooter for the Nestlé corporation. At some stage he decided to chuck it all in and settle in Ayacucho. This decision was, I fancy, helped by meeting a beautiful Ayacuchan who later became his wife. Her father was the mayor of the town. They were all extremely friendly and treated us more like family than temporary interlopers who were filling their tidy milieu with all sorts of film contraptions. He certainly looked after his visitors. I watched him many times meeting new travellers, instantly making them feel at home, and taking an interest in their lives.

Jorge looked like an athletically-built city-slicker who liked dressing up in jungle clothes. His appearance conjured up images of polo players and balmy tropical nights. He would come over to the camp accommodation each evening immaculately dressed in khaki fatigues, knee-high, soft leather boots, with a rum and coke in his hand and tell jungle tales to the wide-eyed tourists. Top that with a sexy accent and pidgin English, it's no wonder some of his clients tried to ... well, 'scrooh him' Julio chipped in.

Julio, who despite his manner was an experienced guide, was employed by Jorge to take care of the three of us for the duration of our stay. He did that admirably. It was in Jorge's camp that we unloaded all our equipment and spent the first and last four weeks of the expedition building the specialised filming sets for close-up photography. It was also here on our first day – 8 September, 1991 – that I was reminded that things do not always go to plan.

The first problem was the rainy season. It usually took place between February and August, but was clearly overrunning by a month. This caused us no end of trouble. Our building of the film set was held up and, because of the poor light, we only had three hours of filming time each day.

The second problem was a budgeting hiccup. Wires, or rather faxes, had somehow become crossed and Jorge was under the impression that we were only staying with him for one month, not two. When I was asked for double the amount of money we hit a major difficulty. There was no way that such a drastic increase in expenditure – $80 per person per day, totalling some $7,500 for the month, would be sanctioned by Survival. I knew there were only two possible solutions. The first was to go home, and the second was for Jorge to give us the second month free. Over a late night game of dominoes and many of Jorge's special '*Cuba libres*' (a rum and coke mix), to my surprise and everlasting gratitude, he waived the extra costs. I raised my glass to him and toasted, 'free Cuba,' to which he replied, 'bugger Cuba.'

To be honest, we all felt a bit like cheats for the first two weeks in the camp. After all, the place was for tourists who were paying a lot of money; for us it was a life of luxury we rarely experienced while filming. Jorge had an arrangement with a local restaurant where we could go every night and eat, drink and be merry. Sheherezade was a dark and dingy eatery with enormous, cheap-looking prints of Lebanese city scenes adorning the walls. A curtain separated the men's urinal from the dining tables. The owner, a friendly Lebanese, always greeted us with the words, 'Good night Americans.' We enjoyed eating there at first but it didn't take us long to realise that almost everything on the menu was a variation on the ham and cheese theme. Wrapped in it, stuffed in it, layered with it or simply sandwiched, the taste of *Queso y Presunto* permeated everything.

Some nights I enjoyed the diversion of 'poowel' with Julio at his friend's pool house. The room smelt of gorgonzola cheese; nicotine-caked paint peeled off the walls, and the floor was sticky with spilt alcohol and spit. The tables were the only thing in good condition – pool was taken very seriously. A man wearing white gloves roved the seven tables, setting up the balls and providing refreshments; he was the only person allowed to chalk the wins and losses on the master blackboard. When you wanted his attention, you had to shout '*piña*' very loudly. I don't know why, because *piña* means pineapple, but if you didn't shout it with great ceremony and volume he wouldn't take any notice. Every game was held for stakes – whether for money or rounds of beer. It would have been very silly to have messed about with *piña* because he looked like a Latin 'odd job' from the Bond film *Goldfinger*.

On one night, playing in pairs with two local men, Julio and I won four straight games and downed the same amount of drinks each. Our opponents, who looked like bank robbers, wouldn't let us go until they had had a chance to level the scores. After another three beers I started missing easy shots. Julio had disappeared into the toilets when it was his turn to play. I refused to take Julio's shot for him and one of our opponents became a little aggressive towards me. He grabbed my arm and tried to push me towards the table. I wrenched my arm free, beating a hasty retreat to the gents to find my partner. I beat on the door of the only occupied cubicle for a joke and said in Spanish, 'I'm the owner, come out.' I was stifling a drunken giggle when the door opened and a big bruiser of a man grabbed me by the front of my T-shirt and pushed me back against the wall. I hadn't realised there were two entrances to the gents. Just as I thought this Goliath Ayacuchan was going to put my lights out, Julio came through the door. He said something to calm the man down and before we had left the stinking urinal the chap slapped me on the back. Julio later told me he had explained to the bruiser that I was a 'stoopid gay gringo looking for his boyfriend'. I never visited the place again.

Some cynic once said that when one door shuts another slams closed right behind it. This is exactly what happened to us the day before we were due to travel into the forest to stay, for the first time, with a community of Piaroa Indians. We were filming some close-ups of a tarantula when Jorge appeared looking glum. Out of the blue, the authorities had announced that no one would be allowed to visit the Piaroa territory until further notice. No reasons were given.

We found out later that the Indians were in dispute with the state and, to put pressure on them, the authorities were effectively cutting them off from trading with the outside world. I was devastated. To be denied entry at this late stage when we had set everything up with such care was deeply disappointing. Our schedule didn't allow for much spare time and the project had already cost a small fortune, with only two rolls of film to show to date.

After long debates with Jorge and his family, I decided to charter a light aircraft the following morning so that we could film the aerial landscape sequences and, at the same time, try to find another Piaroa community in the area to visit.

The view of the rainforest from the air was stunning, although I was still preoccupied with finding a community to film – I can be a bit of a worrier when things don't go smoothly. Rick and I spent several bumpy hours following river courses looking for a likely village. One of the great benefits of film budgets is that they include money for aerial reconnaissance and filming. However, this was an unscheduled sortie and at £500 an hour I was concerned about the cost.

I saw three villages that would fit the bill, assuming the Indians themselves would agree to let us stay with them. We tried to land at a fourth large community that had a rough air strip cut into the forest edge, but some annoyed looking missionaries and two armed helpers waved us away. I marked the three communities on our chart. We now had to travel to these villages by river and ask the Indians for permission to film. The next day, Jorge agreed to put his other work aside and organise a river trip to find the villages. That, it turned out, was typical Jorge. When the chips were down, he pulled out all the stops to help.

My first view of the sun rising over the Amazonas rainforest in Venezuela, South America.

Filming from the plane: dark green forest stretched out below me as far as the eye could see. The magnificent pinnacle of rock or tepuy rising almost vertically from the rainforest is the sacred mountain which the Piaroa Indians call Wahari Kuawai.

The remote Piaroa valley where we would make our film.

A Piaroa longhouse: inside these houses, several families share their lives and possessions - privacy is a concept that doesn't exist. INSETS: (*Left*) *Piaroa hunters are skilled in using blowpipes from an early age.* (*Right*) *A welcome drink of water from a rolled-up vine leaf!*

A Piaroa medicine man, or shaman, starting the ritual for the tarantula hunt.

The creature itself - a dead giant tarantula (Theraphosa blondi).

The special tarantula headdress the medicine man wears before the tarantula hunt.

Filming the Indians catching tarantulas. INSET: *The team photograph! From left to right: world tarantula authority, Rick West; me; Julio, our guide; and my long-suffering assistant, Gordon Buchanan.*

ABOVE *A tarantula in typical defence position.*

RIGHT *The focus of our expedition – a giant tarantula. This surprisingly furry creature has a leg span of 10 inches and a body weight of 4oz. This breed has special hairs that cover its abdomen. When it feels threatened, it rubs its hind legs across the abdomen and the resulting friction sends a cloud of microscopic barbed hairs into the air above it. These hairs are extremely irritating to humans.*

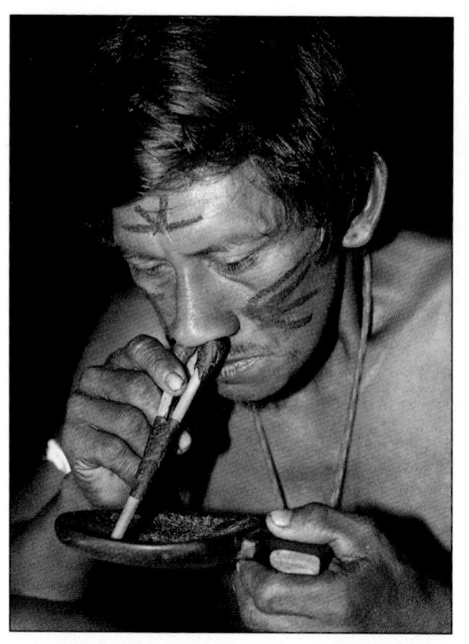

LEFT *The rituals leading up to the hunt: the shaman first tries to raise himself to a higher level of consciousness by snorting the drug yoppo.*

ABOVE *Paying the price for filming the yoppo ritual – I have to snort it too!*

The cave with the Piaroa funeral holes. The walls were adorned with 3,000-year-old scenes. With the bones of their people in front of us, the heat and the strangeness was overwhelming.

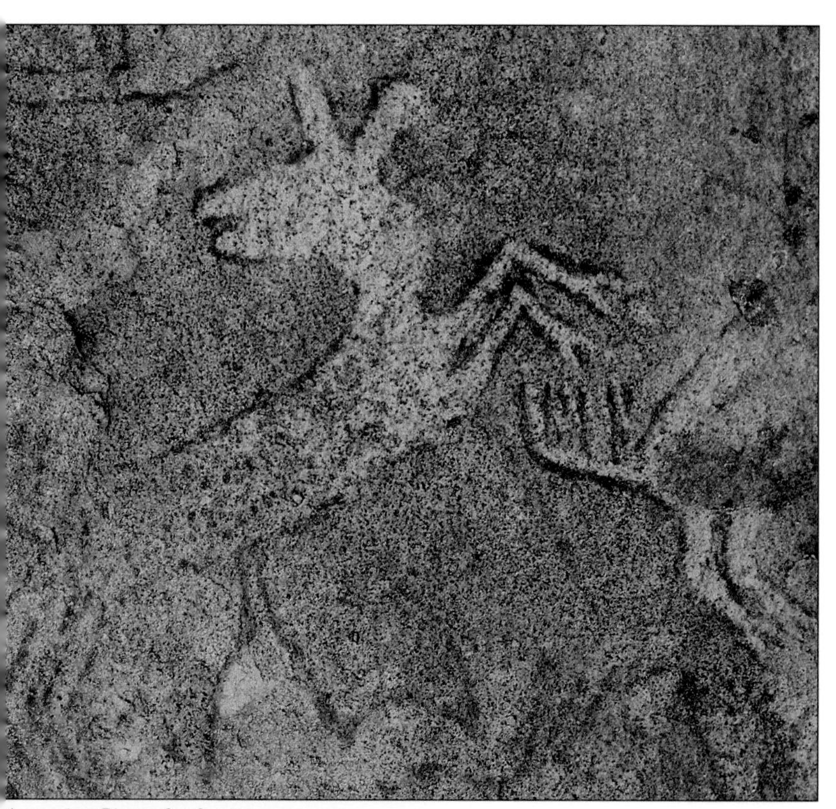

An ancient Piaroa deer-hunting scene.

LEFT *Making a fire by rubbing two pieces of wood together.*

BELOW *A Piaroa prepares the dead tarantula for cooking. Lunch is about to be served. The tarantula's two fangs were removed, the large abdomen twisted off and the contents squeezed out onto a leaf. The leaf was then wrapped parcel-form and placed in the hot ashes at the edge of the fire.*

ABOVE *A fer-de-lance* (Bothrops atrox) *and tarantula in battle. In less than two minutes, the fer-de-lance was dead and seventeen hours later, only its skin remained.*

RIGHT *In the blink of an eye this tarantula jumped – and found refuge in my armpit.*

Chapter Four

Piaroa, Pegleg and the Parasites

We rose early to get to the river before the race between temperature and humidity had began. Humidity was off to a good start before dawn and we were drenched in sweat before we had got into the boat. To make faster headway, we decided that just four of us should make the journey: Jorge, Julio, myself and a Piaroa river guide nicknamed, in Spanish, Pegleg. As a child, he had been spiked in the foot by a venomous stingray, and left disabled with a heavy limp. He seemed not to mind everyone calling him 'Pegleg'.

We had a clapped-out four-wheel-drive to tow our boat along the dusty red earth road to Samariapo, the southern limit of the Orinoco's unnavigable rapids. From there we could travel south by river. As the first glimmer of dawn showed above the tree line, we reached Samariapo and slipped the boat into the river. I was cheered to see undisturbed forest along the river's edge. As the speedboat slid off the trailer, Pegleg fell out, going over the edge of a rock and disappearing underwater for a second before surfacing, spluttering for air and laughing.

By midday, we were belting along an offshoot of the Orinoco and then into a tributary, the Sipapo river. It wasn't long before we came across the first community I had seen from the aircraft the previous day. Jorge and Pegleg clambered up the muddy bank to meet the chief, returning much too soon for my liking. It was bad news; they couldn't let us visit because they were all too busy in their forest farms, or *kanukus* as they are known in the Piaroa language. I bit my lip and refrained from being pushy.

Arriving at the second village, Pegleg went up the steep riverbank, alone. He knew someone who lived in the community and didn't think it would be a problem for us to sleep there. After 15 minutes, with only the light from Julio's cigarette to see by, our emissary returned with a mixture of bad and good news. We could stay the night but not remain or return to film until we had the permission of the chief. He was away fishing for several days and we simply did not have the time to wait for him to come back.

The Piaroa in this village seemed to be friendly and not in any way bothered by our arrival. Pegleg even knew some of them. As the men squatted around a fire boiling coffee, I noticed how small they all were, not one of them more than about five feet four inches tall. I left Julio, Jorge and Pegleg to socialise, with Pegleg, being a Piaroa, interpreting. I slung my hammock under a thatch canopy where the villagers kept pigs – the only available accommodation. The smell was awful. Listening to the others chatting, I buried my head under my sheet and thought of my clean, comfortable bedroom back in Scotland. I finally fell asleep.

At dawn, with smoke drifting lazily from the hut fires, we crawled out from under our thatched canopy. Julio boiled coffee for us, or rather, boiled sugar and added a little coffee to it. Give me black tea any day. Pegleg then brought the *shaman* (medicine man) to meet me. His Spanish name was Luiz, and he was about 35 years old, although it is very hard to judge because living and working in this climate takes its toll on the skin early in life and the Indians always look much older than they are. The *shaman*'s grizzled features clearly showed the hardships of forest life, but he had an open smile. He had learnt Spanish from missionaries and could speak it well. We conversed with Julio interpreting where necessary.

The first and most obvious thing about Luiz was his lower right leg, or more to the point, the lack of it. He had fashioned a rough wooden leg from forest wood and strapped it to his thigh.

Two years earlier, Luiz had been hunting in the forest close to his village. It was just before dawn and there was enough light to silhouette a monkey in the canopy. He spotted a group of howler monkeys and crept closer to get a better shot at one with his blowpipe and poison darts. Suddenly he felt a pain in his shin and jumped with shock, realising instantly that he had been bitten by a large *fer de lance*. He had accidentally stepped on the snake mid-body and as it swung around to defend itself, it sank its fangs into the skin above Luiz's ankle.

He told me that, without hesitating, he had cut off his leg below the knee with a machete. Machetes are prized possessions and all forest people keep them razor sharp, but I couldn't imagine how a man summoned up the strength of mind to be capable of cutting his own leg off and remain conscious. He said there wasn't time to get drunk because the snake bite kills so quickly. I asked him how he treated the open wound and he told me that they use forest cures. In this case he had caked his severed stump with a mixture of earth and ground tree bark. Julio looked as though he was going to throw up, but he lit up another cigarette instead.

I chatted with Luiz about my own near-misses with snakes – tame compared to his – and asked if I could take a photograph of him with my compact camera to show my little girl. He agreed. The photo never came out properly but I kept it as a poor record of a very remarkable man.

During the evening with Luiz and his people we learnt of another small forest village some nine miles away where a relative of one of our hosts lived. He was sure that we would be able to film there. The villagers there were all familiar with the giant tarantula, their people ate it, and their medicine men prayed to its spirits. We decided to set off the following day, and after eating a dinner of paca and rice, slept soundly under the thatched canopy of their manioc hut.

The morning was heavily overcast, matching my mood. At least we had a hopeful introduction to the third community. As we set off, Jorge promised Luiz that he would put on a special show for the village when we passed them on our return down river, hopefully later that same day. I asked him what he was talking about but he was in a mood to play games and told me to wait and see.

I soon forgot about it as I took in the scenery and scanned the forest for birds. This northern quarter of one of the last greatest stretches of unspoiled forest was simply beautiful. Each bend in the river gave us different sights. With cloud cover breaking up quickly, allowing the sunlight to filter through, flowering trees were dripping their brilliant yellow blossoms onto the black river. A huge fish jumped clear of the water and Jorge shouted '*Aruwana*', its local name that literally means 'jumping fish'. The drape of foliage was unbroken, a curtain of continuous vegetation.

After a further hour at speed, we rounded one curve in the river and there was my first sight from the ground of the sacred mountain, Wahari Kuawai. Some mountains, like some waterfalls, move me to silence. They have an immensely powerful presence and it is easy to understand why many indigenous people create folklore around them. For me they are evidence of the earth's hidden powers, and at least one thing that mankind cannot mess with. They seem to put us humans in our place and make us realise how insignificant we are compared to them. Wahari Kuawai is firmly on my 'special places I have seen' list.

This part of the country belongs to the oldest geological part of Venezuela and is one of the oldest landforms on the South American continent. The sedimentary rocks, mainly sandstone, are thousands of feet thick. The underlying rock foundations are estimated to be 1,600 to 2,000 million years old. The whole area has been subjected to periods of massive uplift and erosion, and the erosion of the sandstone deposits ultimately left these isolated dramatic mountain forms or *tepuys* with almost vertical rock walls. From some of these flow spectacular waterfalls, the most famous of which is Angel Falls, north-east of Wahari Kuawai, which cascades 3,000 feet to the forest below.

We kept snatching tantalising glimpses of Wahari Kuawai as the river began to wind more severely. Jorge brought me to my senses when he announced, 'looks like the place!'. I could just about make out the tops of thatched roofs about 500 yards ahead. I grabbed the film camera from its case and had taken about 30 seconds of film of the village from river level when Jorge told me to put the camera away before we got too close. Piaroa are highly sensitive to intrusion and I could have messed up our chances before we had arrived. I just couldn't help myself – it was such a lovely sight. That first shot did make it into the film though.

Our boat pulled into the bank at the bottom of a sturdy wooden ladder that rose almost vertically up the 30-foot-high bank. The rungs were about 18 inches apart and I remember thinking it would be some climb with the camera gear. A small naked child, an older man and a scraggy-looking dog were peering down at us. I smiled at them. The child and dog withdrew quickly. Jorge and Pegleg decided to go up alone first and said they would call me if things looked promising. It was a long ten-minute wait before I saw Jorge coming backwards down the ladder. He was alone and I thought the worst. He turned around at the bottom of the ladder, a big grin on his face. They wanted to meet the foreigner and see his camera machine.

We struggled and panted as we hauled the heavy camera case and tripod up the ladder. Reaching the top, the first thing I noticed was an incongruous huge solar-energy panel on stilts in the space between three of the thatched houses. I shook hands warmly with two Piaroa Indians. No one else was in sight, not even the mangy dog. I took the camera out of its box and set it up on the tripod for them to look through, pointing it down the river. As the first man looked through it I zoomed the lens to its maximum and the man made a sound that obviously meant 'wow'! His friend pushed him to one side and I zoomed the lens again for him. After five minutes of this, other men and a few children began to gather. There were still no women, though. This turned out to be a common trait in Piaroa villages; the women were shy and also, it transpired, they did most of the hard work in the forest.

Piaroa architecture is strikingly different from any other that I have seen in the rainforest. There are two basic buildings, the rectangular long-houses and *churuatas,* large dome-shaped structures some 100 feet in diameter and 50 feet high. Both types of building are constructed from forest hardwood frames and covered in a thick layer of manaca palm thatch. This community had one *churuata* and six long-houses, where 90 men, women and children, and their dogs, chickens and various forest pets lived. Inside these houses many different families shared their lives and possessions. Privacy was a concept that didn't exist.

I particularly liked the *churuata* because of its unusual shape and design. It has a pointed summit where, usually, a bare, spindly, forked branch poked through. This, so the Piaroa told me, gave the dwelling an earthly link to the invisible spiritual world of their heavens. Inside this dome-shaped house as many as 40 people may live. We were lucky to see a *churuata* in the stages of being built and could only marvel at the technical skills employed in its design and in intertwining the many lateral support timbers. The covering was made by knitting thousands of palm fronds together. The entrances are about four feet six inches high and anybody except a small child would have to stoop to get in. The buildings are dark inside, but eyes adapt quickly and gradually things begin to appear and take shape. At night the interior is lit by fires and small lamps that burn on animal fat.

The first time I entered a *churuata* it felt very homely with hammocks slung around the edge and small fires smouldering between them. For the first time, I also saw a few women and children. Two young mothers were swinging in hammocks breast-feeding their babies, while three young children chased a tame agouti, a forest rodent that looks like a cross between a squirrel and a rabbit. The children ran around the dirt floor after it until one of them rugby-tackled it and then cuddled it for half a minute before becoming bored and letting it go. Then the chase started all over again.

In another corner, an ancient-looking man was making a hammock out of vines while he chewed what looked like a big wad of black tobacco. He spat out a mouthful of thick goo on to the floor. I noticed that all the toes on his left foot were missing. It looked like leprosy. The overriding feeling was one of harmony, although I recently received a letter from my parasitologist chum, Dr Guy Barnish, from the Liverpool School of Tropical Medicine (or 'Trop Shop' as he calls it), which never allowed me to view these attractive dwellings in quite the same way again.

A parasite is defined as an animal or plant living in or on another animal or plant. Guy spends his life travelling around the world studying the animal varieties, usually collecting his samples from the faeces of the natives. Guy, and his wife Simone, had invited me to a barbecue at their home before I left for Venezuela. In their back garden on that late summer evening in August, as I savoured every mouthful of food, Guy said, 'You've got a perfect body for parasites, Nick!'. He continued, 'Your body maintains a constant internal temperature of 37 degrees centigrade; inside it is always warm, dark, with plenty of nutrients.' I took a bite from my herb sausage smothered in HP Sauce while Guy went on. 'On the outside there is always some body hair sprouting through the thin layer of your skin – in fact Nick, your body is a stable environment for a staggering number of organisms.'

'Bank managers and tax inspectors included,' I replied.

'They are living on or in you now matey.'

'Better feed them then, Guy.' Swallowing the last bit of mouth-watering sausage, I asked, 'So what can I expect to find in Venezuela?'

'I'll do some research this week and send the information to you.'

This letter arrived several weeks later.

22 October 1991

Dear Nick,

Wherever will they send you next? Anyway, it seems hot-foot from Tiwai Island in the middle of the Moa River of Sierra Leone you are to be dumped in the hinterland of Venezuela. Well, I suppose it's one way to see a part of the world that tourists don't reach.

Well, my friend, as promised I have done some research into what may be waiting in the shadows ready to pounce. There are indeed a few parasites of which you may become the unwitting host!

As I have told you before, your lifestyle, while on location, puts you at risk of entertaining quite a few infamous, and not too photogenic, little beasts. Generally, the parasites will get to, or into, you either from close contact with other people, through your drink, food, or by way of your skin.

If you get in close contact with people for long enough you could get three kinds of lice: head, body or pubic, and the scabies mite. The first three live on the dry skin of your head, body or pubic area, cause profuse itching and breed like mad. You could catch louse-borne typhus fever from the body louse. The scabies mites tend to live in the folds of your skin, such as between fingers or in the elbow crease – but other places also become des reses, as long as they can burrow in your skin and make a comfortable living off you. They also itch like mad.

Cleanliness is next to Godliness as far as keeping that little cluster away from you, so do bathe as often as you can. Of course, you will almost certainly pick up ticks as you meander through the vegetation – there is always the potential to pick up something nasty from them. Be careful how you pull them off or you could leave their biting parts in your skin, and then you will get a nasty secondary infection which will require antibiotics to cure. I tell you, it's tough in the tropics!

What else can you get out there? Biting flies! They are bad enough, but when they leave a minute visitor or two in your blood or skin, then you really have to start diagnosing and treating. The obvious one that comes to mind is the malaria parasite – transmitted by mosquitoes – a tiny little thing which lives (for much of its life) in the red blood cells. It doesn't just sit there though, it reproduces and then a myriad of progeny burst out to reinfect the red blood cells. When all the cells burst at the same time you go down with the malaria fever, but you know all about that!

Of course the mozzies bite at night, but where you are – and I bet you have already discovered this – there are day-biting flies. You recall the blackflies of the Moa River in West Africa? Well they are there in South America too – I think they call them *piums* out there and, as in Africa, they transmit *Onchocerciasis* or river blindness. The small worms (*microfilariae*) pass into your skin from the blackfly while it is feeding on your blood. They wander around inside you, grow up and eventually the female worm is 20 inches long!! The male is much shorter, but the two wrap themselves up into a ball just under the surface of the skin and produce thousands of babies. They all set off and wriggle through your skin just waiting to be picked up by another blackfly. The little worms sometimes wander through the eye, and if they stop there, then the trouble really starts and the end result is blindness of one sort or another.

There's another tiny biting fly called *Lutzomyia.* A most delicate little thing with lovely fringed wings, but if that is infective, it gives you a dose of *mucocutaneous leishmaniasis,* and this you certainly do not want! Ulcers develop in your mouth and nose region and eat away your soft palate, eventually totally disfiguring your rugged good looks. Their reservoir hosts include sloths and anteaters – just the sort of animals you'll be wanting to film.

The largest biting insects you will encounter are *reduviids* – what my kids would call 'gross'. One of its common names is the 'Kissing Bug', and as a biologist you will appreciate that this one is an hemipteran bug. It comes out of cracks in mud and thatched walls of native dwellings at night and pierces the skin with its fine proboscis. Then it sucks up your blood until it looks like a small grape! What a kiss! In fact, and this must interest a beer drinker, as it fills up with blood at one end, it gets the urge to relieve itself from the other end. Naturally, the act of taking a swift nip causes some irritation and as you rub or scratch the cause of the itch you may well rub some of its excreta into the bite.

If the Kissing Bug was infective, then you introduce some other small parasite into your blood. This time it is Chaga's disease or South American *tripanosomiasis* – there is some evidence that Darwin suffered from this after being bitten by bugs in Mendosa, Argentina. Still, that's another story. The result of this infection is that you may suffer from a grossly enlarged colon, heart, or both! Beware Nick, some of these bugs like to feed on domestic animals, and also the type of orphaned forest animals that you keep in your menagerie. Pick them up with tweezers, not your fingers – that way you will keep their excreta away from grazes and cuts.

Onchocerciasis, leishmaniasis and *trypanosomiasis* all take years to develop, so unless you're going to be there for a long time, you may not even realise that you have picked up any of these.

The rest of the parasitic fauna that you might get are water or food borne. What's the most common problem people from temperate climates get when they go to the tropics? I know you have had it plenty of times, I have, too! Diarrhoea and its ally dysentery. Montezuma's Revenge, Delhi

Belly, Gyppo Guts – they're all the same as far as the poor sod who's got the squits is concerned. Basically diarrhoea (funny spelling isn't it?) is when you have the runs with mucous, and dysentery is when you pass six or more blood-flecked stools a day. The cause is either a protozoan called *Entamoeba histolytica,* a variety of bacteria or a load of intestinal worms. Boil your water before you drink it and, as I have always told you, brush your teeth in beer! I could do with one now, couldn't you?

There are about 13 little single-celled animals which can happily live in the warmth of your gut. Generally you will hardly notice them, but one makes you fart quite a bit – that's *Giardia.* It looks like a pear sliced in half lengthways and has four pairs of long hairs – *flagellae.* The flat side acts as a sucker and it clamps itself onto your gut wall and feeds. As I said, it causes flatulence and, if you have plenty of them, malabsorption of fats, so your shit floats – good one, eh? Needless to say all these little beasts are transmitted via the faecal-oral route, so wash your hands after you've been to the loo, and watch out for those using your river water as a bathroom/toilet. Flies are quite good at carrying the resistant cysts from exposed crap to your food, so swat the buggers before they land! Should keep you busy.

What else is there? Ah-ha? Worms. Roundworms, whipworms, hookworms, tapeworms and flatworms. They all find a nice warm home in various parts of your gut, bathed in food and protected from enemies – it's no surprise that they like it there.

Mention Deuteronomy,
SURVIVAL of the fittest!
Best of British,
As ever, GUY.

When we emerged from the *churuata* into the bright sunlight, the contrast was so great that I had to shield my face with both hands for my eyes to readjust. There was the solar panel again about as incongruous to my eyes as a block of flats on Mars. A scarlet macaw was now perched on one corner of it. There were no cables running from it so it could not have been in use. Where had it come from? The older of the two Piaroa men explained how some missionaries from a group called the New Tribe had brought it a few months ago and would be returning with light bulbs for them to be able to see at night. To reinforce the incongruity, my eyes were drawn to the outside of the round house, where there were about 20 white discs, the size of dustbin lids, which one of the Piaroa was flopping over one by one to dry in the sun. They were cassava breads made from the manioc plant (also the source of tapioca), and a staple diet of the Piaroa.

Our prospects for filming were looking good. The Piaroa who had been talking about the missionaries said we should follow him to the *kanuku*, their forest farm about 15 minutes' walk away. *Kanukus* are mainly a communal form of agriculture, although some Piaroa have additional patches for their family to cultivate.

The first impression one gets walking through a *kanuku* is how well organised and tended it is. This one had layers of banana, cashew and jambo trees (producing a bright red apple-flavoured fruit) arranged to shade smaller plants which would die if left in direct sunlight. The second impression is of the great variety of fruits and vegetables that are grown successfully in the notoriously poor rainforest soils. Among the crops, I saw manioc, bananas, passion fruit, hot and sweet peppers, various palm fruits, maize, coffee, pineapples and sweet potatoes. There was even a plant whose ripe fruits were harvested and used for fishing, but not as bait for catching vegetarian piranha! The crushed fruit pulp is a highly toxic substance whose poison paralyses the gills of the fish. Its effect, however, only lasts for a few minutes and if not gathered in quickly the fish recover and swim away. When the Indians cast the paste into a stretch of creek or river, dammed to prevent escape, the fish float to the surface where they can be easily collected. Cooking the fish destroys any harmful poison residue in the meat.

As we walked back to the houses I couldn't help asking why these people weren't left in peace. The solar panel between the huts was just the thin end of the wedge. I accepted that the Piaroa wanted machetes and cook pots and fishing hooks and all the other useful modern conveniences, and that they had every right to have them. But these things can, and do, reach them through river trading in a natural way. I felt that the politicians who have such a big influence over them should ensure that 'progress' is more sympathetic to the Piaroa' way of life. So often we met Indians who had become displaced alcoholics, wandering lost and penniless in so-called civilised conurbations, like Puerto Ayacucho.

The walk through the Piaroa's *kanuku* was enlightening and I was beginning to feel reassured about the film we were going to get. Then Julio dropped a verbal bomb. We weren't going to be able to stay more than a couple of days because all the Piaroa men would be going onto the farm to cultivate it. I couldn't believe what was happening – once again, it was one step forwards and two back. Jorge elaborated on what Julio had told me, adding that the Piaroa had told him of a smaller community further north where the men had finished their planting, and where we would be able to film them hunting.

Despite Gordon being back in Ayacucho I decided to start filming at our present village immediately. Two days were better than none at this point. With Jorge and Julio pressganged into being film crew, we tried to capture the way the Piaroa lived and farmed. Jorge managed almost to sever one of my fingers setting up the tripod and Julio nearly split his pot belly laughing. At least I was getting something useful in the can. Towards the end of the second day, as we were leaving, I fell down the last ten feet of ladder and landed unceremoniously in the deep mud at the river's edge. Unlike when we arrived, there were at least 20 Piaroa there to wave us off and see me drop. Caked with thick mud, I picked myself up with as much dignity that the pain in my leg would allow and smiled, pretending to laugh through gritted teeth. They were all laughing, too, but not as much as Julio.

As we retraced our river journey to Ayacucho, I decided that the tarantula expedition would have to be delayed again. It was imperative to start working with Rick in the film sets. Jorge, forever optimistic, said he would have everything fixed up for us to film at a place called Punto Bravo in two weeks' time. I thought it would be better named 'Last Chanceville'.

Jorge could never pronounce the word south properly; he always said 'thous' as in, 'We go thous for over an hour now – time to stop.' Why, I asked? He told me he wanted to water ski for a while and my first reaction was how bizarre, but if he wants to, why not? The two skis were stashed below the boat's seats. I took the helm and after ten minutes with Jorge smiling and giving me signs to go faster and faster, we came around a corner. Only then did I understand what all of this was about. Luiz, the one-legged *shaman,* and a crowd of about 60 Piaroa had heard the approaching engine and were spread along the riverbank in front of their houses watching as Jorge waved and then fell flat on his face. Despite three attempts I couldn't pull him up again and finally had to tow him into the village canoe landing.

Luiz was there with a beautiful festive parrot on his shoulder. With his wooden leg he looked like a Piaroa Long Juan Silver. Then Jorge, in his inimitable style, clambered out exclaiming 'o sheet my wallet'. Typically, he had forgotten it was in his pocket when he leapt over the side of the boat to ski. He spread his money and identity papers, driving licence and other oddments on top of the speedboat's bow to dry and we all sat about enjoying the afternoon. A cooling breeze from the river brought the scent of a nearby flowering tree. I breathed in deeply, a million miles away from pressure and at peace with myself.

Looking at Luiz's parrot sparked off a long talk with Julio and Luiz about the different forest animals his people kept as pets. The most common were parrots, macaws and toucans. Sometimes they would collect a baby monkey that had fallen with its parents during a hunting session. These 'pets' do become extremely tame. The year before they had even raised a harpy eagle, though they had eventually had to kill it because it was taking their chickens.

Luiz disappeared, and returned with his blowpipe, quiver of darts and a gourd filled with the poison *curare.* I was admiring the pipe when Luiz muttered something to Jorge. Jorge told me that Luiz wanted to give me the weapon as a gift, in memory of him. I was touched by the gesture but tried to refuse because I knew these weapons took so long to make and are essential to the Piaroa's survival. But Luiz ignored my protests and presented the blowpipe to me, with the darts and gourd. They now hang proudly on my lounge wall in Scotland as a reminder of the experience. Thank heavens Jorge had fallen off his skis.

We left Luiz and the Piaroa in the late afternoon, feeling sad that they were not to be the community we would film. But having spent time with them hardened my resolve to, somehow, record their lifestyle.

Chapter Five

First Catch Your Tarantula

On 15 November, 1991 we set off for Punto Bravo – the brave mountain – in the hope of filming a tarantula hunt and the associated Piaroa rituals. We had met a friendly, crazy Brit called Lee who was travelling in the area and who had stopped off at Jorge's camp for a few days. He told me that Survival films were his favourite television programmes and asked if he could tag along for a ride to Turiba, 150 miles north of Ayacucho. Gordon noticed that his rucksack seemed to contain nothing but tins of beer so we felt we couldn't refuse.

We left Jorge's camp at sunset with our two four-wheel drive vehicles. Gordon, Lee and myself were in the first, with Julio driving. Following us were Rick Manolito – a young Venezuelan boy hired as cook – and a sour-faced driver whose name I never discovered. The journey north through the night was monotonous and, with Lee's beer thinning our blood, cold.

A rough bush track swallowed us and our vehicle, and the little light we enjoyed from the equatorial night sky vanished.

We were only 15 or 20 minutes into the dark tangle of vegetation when the first creek appeared. Just minutes later we were stuck in its flow, the Land Rover's wheels slipping on the bedrock of the river, with the surging waters at bonnet level. Gordon – who for some reason was whistling *Flower of Scotland* – and I tried to push the vehicle free but the water was running so fast we were in danger of being swept away. It took half an hour and some good driving sense from Julio to extricate us. The driver of the second Land Rover refused at first to cross the creek. His vehicle was filled to the roof with our expensive and fragile equipment. Equally as precious, of course, was Rick, who was perched in the back. We resorted to bribing the driver with money and, with us shouting encouragement, he finally made it across. Then we bounced over rocks and fallen wood, eventually climbing up and out of the forest into an area of scrub. It was here that we would make camp. In the glimmer of our one working headlamp, I could make out thatched roofs. It was the Piaroa settlement.

The sound of our approach roused a few Piaroas from their slumber. They crept forward cautiously, peering around corners. I saw the women pulling some children back as though we might devour them. Several dogs barked until someone kicked one and they all ran away howling. In time, considering we couldn't speak their language, most of those barriers were to melt and we came to know them as well as one could under such circumstances.

As we were unloading our two Land Rovers and putting our equipment boxes into one of their store huts, I felt my back give way. An occupational hazard! I winced with the pain and sat on one of the camera boxes. I have been plagued with lower back problems for a few years – a common price paid for a career in this business. I knew that the following morning I would be in agony.

I delved into my rucksack and pulled out two water-colour paintings my daughter Emma had given me. I missed her so much. I was going to put her paintings on one of the hut supports when I became aware of two tiny faces watching me. Like children all over the world, they were filled with curiosity. I beckoned with my hand for them to come closer and after a few minutes of cajoling they were by my side, touching my arm to see what it felt like. I gave one painting to each of them and they smiled openly. I asked Julio to take a picture of me with the girls holding Emma's paintings up so that I could show it to Emma when I eventually went home. I was sure those two pieces of paper were going to adorn the walls of their homes until eaten by the hordes of ravenous insects that share their dwellings.

We were given an old manioc store for our stay. It was mud-walled with a thatch covering and we strung our hammocks on a twisted rafter. The place was stuffy and infested with thousands of cockroaches. They didn't waste any time in colonising every nook and cranny of the equipment and supply cases. They are the only insects I detest, especially their awful habit of scuttling across your face in the night. Heaven help you if you snore or sleep with your mouth open.

One night Gordon dreamt he was eating crispy onion rings and woke in a startled panic. He had bitten a large roach in half and swallowed it! His shout of disgust woke me and in the light of my failing torch batteries I could see some of the creature's horrible white liquid dribbling down his face. We still managed a laugh.

Gordon and I shared the manioc hut but Julio decided to sleep in one of the Land Rovers 'for our security,' he told us – or, more likely, we muttered cynically, just to escape the roaches.

Having done this kind of work for so long, we tend to be a little blasé and a situation has to be pretty dire for us to feel threatened. It is only when others less hardened to such experiences, like Julio, show concern that I realise we might be a little exposed. I like to believe Julio slept in the car because he didn't want to take any chances with our precious supplies of food and that, it turned out, was a wise decision. When we unloaded our cargo into the hut, a crowd of Piaroa began to gather and it was obvious that the sight of our supplies was causing excitement. Cases of tuna fish, peas, sweetcorn, cooking oil, corned beef, beer tins, tinned tomatoes, toilet paper and other luxuries that offered a change from eating monkey or similar meat were examined closely. The Piaroa's chatter rose in volume until the chief appeared and sent everyone off with some harsh words that none of us could understand.

'This buhster thinks yoove bort this sturf for him,' Julio pronounced. The chief then asked us for sugar, coffee, powdered milk and toilet paper. My first impression of him was not good. It was indefinable but I begrudged him even one roll of toilet tissue, saying to Rick, 'Why can't he use leaves like he always does?'

The chief could speak enough Spanish for Julio to be able to communicate with him although there were still many misunderstandings along the way. We hadn't eaten for 12 hours so before we climbed into our hammocks we tucked into a supper of Gordon's tuna delight. Recipe as follows: mix tins of tuna, sweetcorn, peeled tomatoes and mayonnaise and spread with fingers onto dry wheat crackers. For extra flavour sprinkle crushed chilli pepper, as Gordon and I did. Thoughts of home and restaurants were far from our minds as we devoured them hungrily.

It rained heavily in the night and the thatched roof leaked so badly that we had to set up umbrellas to keep us dry. Night visitors were also in evidence and the biting flies, not to mention the cockroaches, were the worst I have ever come across. The mosquitoes flew in squadrons, but the tiny blackfly were even worse. Their bite leaves a tiny speck of blood on the skin that hardens quickly and after a while the irritation can drive you mad. Nothing takes that dreadful itchiness away. After 20 minutes outside the hut, I had more than 150 bite marks on the backs of my arms alone. I am not exaggerating, because I counted just about every one. We knew that, as my friend Dr Barnish had warned me, these little darlings carry *onchocerciasis*. I dread to think what horrors are still lurking, like time bombs, in my body.

The next morning, I was woken by the sound of our cook, Manolito, outside our hut singing Elvis Presley's *Teddy Bear* – but in Spanish. Gordon and I were soon laughing and singing along too. It was a happy start to our first day.

It was fresh and sunny, one of those days when you stretch your arms out and breathe in the air deeply, a great day to be alive. There was a river about a ten-minute walk away where the villagers washed and bathed. The problem was that the path down to the waterside was also where most of the Indians went to defecate. The smell was appalling. Rick didn't mind too much because it attracted many different species of dung beetles. 'Hey guys, come and look at this beauty,' he would shout, while watching the 'beauty' devour a pile of stinking faeces.

Gordon and I didn't share his fascination and went down to the river to wash and clean our teeth. A brief swim and some stretch exercises seemed to relieve the spasmed muscles in my lower back. I used up the last of my shaving foam, knowing how uncomfortable I would feel in a few days as a beard would begin to sprout.

In the warm early morning light even the prospect of Manolito's fly-encrusted breakfast seemed cheering. As the treacle-like coffee came to the boil I saw the chief talking to Julio, looking extremely serious, and I knew something was wrong. Julio beckoned to me to join them. I was wondering what on earth it could be about. It turned out that the chief was furious that I had taken a photograph of the two little girls the previous night without his permission and was demanding compensation. Rick looked aghast.

Julio and I were upset that such a simple act had caused a rift in our relationship right at the start of our visit. There was a deeper concern for me, too, in that it would be impossible to work there if we had to ask the chief permission every time we wanted to film. I voiced this worry through Julio. I also told the chief, again through Julio, that if there was anyone who didn't want us to film or photograph them, we would obviously respect their wishes. The chief then floored me by saying he wouldn't give permission to film anyone. We could stay with them, but definitely *no filming!* I wanted to ask him to return the toilet paper but, joking apart, I knew the only solution was with Jorge, hundreds of miles away, who had conducted the initial negotiations. The frustrations of this work sometimes overwhelm me and we were facing just that situation now.

Jorge was due to visit, but not for another two days. Until then we simply had to wait. Depression descended. Rick was thoroughly upset by the events, not being used to this, whereas it was a familiar scene for Gordon and myself. Rick decided to return to Ayacucho and await our return. There he could at least sate his passion for collecting bugs and, in the meantime, would also try to find us different tarantula specimens to film as well as keeping an eye on our special film sets. Julio managed somehow to persuade the chief to permit me to film the scenery and to go into the mountain area with their hunters, though not to film them. We busied ourselves doing that until Jorge arrived.

Unlike the forest with its canopy of trees, the village had no cooling shelter. The only shade was inside the huts, which were horribly stuffy. During the hot afternoon of our third day we were sitting outside our hut when we heard an approaching engine. The open-top four-wheel-drive vehicle came towards us along the track, billowing clouds of yellow dust behind it, and the hilarious sight of Jorge lifted my battered spirits no end. He was standing on the passenger seat holding onto the windscreen frame with a driver sitting at his side. Jorge was dressed in his usual designer jungle wear, cravat fluttering over his left shoulder and with an Indiana Jones hat. Topping off the spectacle was a huge bowie knife, almost as long as his thigh, which he had strapped to his left leg. He was waving to the Piaroa as he entered the village. I noticed that they were even waving back. It was like something from an old great white hunter movie. 'What a wally,' – well something like that, my companions chorused, and looking at Jorge at that moment I couldn't disagree. Nevertheless I was so happy to see him.

There wasn't much time for small talk over a cup of tea. First we didn't have any tea bags left and, second, Jorge was only staying for a few hours. He had made the 15-hour round trip just to make sure we were settled in all right. I explained what had happened and in true Mr Fix-it fashion Jorge said, 'You no worry Nick.' He then strode across to the chief who was waiting at the back of the gathering crowd. Jorge isn't a tall man but he made the chief look positively tiny. I saw Jorge give him a parcel, which later he told me contained cotton material for the chief's wife, and then they went into the headman's hut and talked for 20 minutes. Eventually Jorge came out alone. As simply as the problem seemed to have started, so it had ended. We could film and photograph unless the individuals concerned did not want us to. Strangely, we never saw the chief again. The one thing I have learnt during my many encounters with indigenous people is that you can never predict what is going to happen tomorrow.

Over the following weeks we grew closest to three Piaroa men in particular – Lanyetsa, Wariena, and the medicine man, Panutsa. We spent our days and nights with them, while they hunted, rested, spoke to their spirits, and did the things that Piaroa do.

Panutsa, it transpired, had been raised as a child in a community contacted by missionaries. He had learnt to speak Spanish well, which allowed us to communicate, with Julio interpreting. They were small men in stature, their average height being about five feet. Their jet black hair always looked as though it had been cut in a salon, with no angular edges or roughly cut layers. We never found out how they managed that.

Wariena looked the youngest, about 28 years old. Unlike the other two who had more rounded faces, his was long and pointed, with two warts on his upper lip. He was the more silent of the three, but was easily led into laughing fits by the other two, especially when the hilarity was prompted by one of us foreigners falling over in the forest. Lanyetsa was a little older and more handsome. He laughed more than the others, wasn't as shy as Wariena and, it turned out, was the easiest to film. He seemed almost to understand what we were doing with our silly machines.

Panutsa was the oldest, between 40 and 45, as well as the tallest. He wore a cloth around his middle, while the other two wore shorts given to them by missionaries. He had whiskers on his chin whereas the other two, like nearly all Piaroa men, were smooth skinned. Their skin was coffee-coloured and bore some scars of old wounds, especially on the lower legs. Panutsa spent so much time drugged up to the eyeballs with *yoppo*, a hallucinogenic powder, that he drifted about the village in a world of his own. His position in the community was strong – the Piaroa place much faith in the magical powers of the medicine men – and little of importance went on without his involvement.

To let us into their lives was indeed a privilege and, I thought, showed enormous trust. But perhaps trust was not an emotion the Piaroa were familiar with because I don't think there was any reason to distrust anybody in their way of life – apart from politicians, of course. For our part we respected them, were genuinely interested in their culture, language and everyday lives, and would not have filmed anything they were unhappy about.

By a stroke of good luck – just to prove it does happen sometimes – they were about to embark upon a hunting trip into the forest and would be looking for the world's largest spider, 'our' *Theraphosa blondi,* to snack upon. For them the spiders are the equivalent of shellfish or caviar.

The rituals leading up to the hunt were fascinating. The *shaman* first has to raise himself onto a higher level of consciousness so that he can communicate with his spirits. This he does by inhaling a powder ground up from the seeds of a forest liana which is called by the same name as the plant, *yoppo.* I spent some time, with Julio interpreting, talking to the *shaman* about this. He told me that under the influence of the drug he sees the spirits of his family, friends and animals. In particular he sees the spirit of the giant tarantula and would pray to it so that it would lay more eggs and eventually provide them with more food. He would also see bad spirits trying to steal the spiders from him which he would have to fight off.

I naturally wanted to film this ritual. To my surprise Panutsa agreed, but with one condition. Julio translated: 'Yoou can film it alll buusterrr, but yoou gotta doo it too.' I thought it was a small price to pay and in a deeper sense only right. I wanted Panutsa to understand that we were sincere and not intruding lightly. I will never know what he really thought about us but I agreed, although I added my own criteria that I was not to take the *yoppo* until after I had filmed him as I wanted to be able to operate the camera! I also told him I wanted Gordon to do it as well – why should I be the only one to suffer? It was agreed.

It was about half-past four, an hour and a half before dawn. The *shaman,* dressed only in his lap cloth, sat on his haunches in the centre of his long-house, the atmosphere heavy with the strange odour of the *yoppo.* A small fire billowed smoke into the air around him, filling the place with eye-watering fumes. At his feet was an odd assembly of artefacts, very strange to my eyes. Julio, Gordon and I watched, learned and filmed. To Panutsa's left lay two hideous-looking masks. One was a monkey's face with a horribly contorted expression, the other a tarantula and extremely lifelike. They were made from wild beeswax. In front of him were a pestle and mortar made of forest hardwood, a small brush and a large leaf spread with some small unidentifiable seeds. A little glass bottle, the only modern item visible, contained some *yoppo* already in powder form.

Hanging from the *shaman*'s left arm was a large lump of quartz. It was, he told us, his lucky rock. His preparations were meticulous. He ground the seeds and brushed the powder on to a small wooden dish the size of a saucer, carefully cleaning the tools he used after each action. His *yoppo* pipe was made from animal bones, probably monkey. It was wishbone-shaped, the single end being put into the powder and the forked end precisely formed to fit up both nostrils.

He drew his breath in slowly but definitely. Seconds later he gagged and a stream of mucous flowed from his nose. He coughed with convulsions. Gordon gave me a knowing look, screwing his face up as he listened to it all in glorious amplified stereophonic sound through the digital recorder's head-phones clamped around his ears. The recording was so horribly graphic that I wondered whether we would be able to use it at prime viewing time in the West when many people would have supper on their laps.

The *shaman* ground up more powder and repeated the whole process another three times. The medicine man's eyes now betrayed his condition. Trancelike he sat there for a long time. I cannot be more precise about that period for it was a contagious atmosphere. All of us were silent, watching, in our own drug-free trances almost disbelieving that we were there witnessing it. It seemed somehow unreal. Where was he?

My stomach felt distinctly queasy and I wished I could renege on the pact. But it was soon my turn. As I squatted next to the *shaman* I told myself that at least we had a remarkable film sequence in the can. I was thinking of England when Gordon took over the camera to film me going through the ritual, proof for the production team back home of the lengths to which we have to go to bring film in for Survival.

The *shaman* ground up more seed and looked at me with what I thought was an odd smile. Julio was sniggering. I did wonder if it were a Piaroa leg-pull and whether in the morning they would all split their sides laughing. The *shaman* shoved the small wooden dish and *yoppo* pipe under my nose and nodded to me. Body language is a wonderful thing. We couldn't converse directly but I knew he was saying 'go on' and I'm damn sure he knew I was saying 'let me get out of here, name your price'. The pipe ends were dripping with his phlegm, and I felt my throat seal up.

The sensations that followed happened in less than a few seconds. My nose caught fire, my sinuses swelled and then exploded. Somewhere deep in my brain someone was drilling and my eyes streamed in sympathy with the *shaman*'s. No higher plane for me, only the earth floor to throw up upon. The *shaman* did laugh, but not in an unpleasant way. He slapped me on my back muttering something that Julio translated as, 'you are his brother now'. I wish I could say that I felt proud, but, alas, only sick to the stomach. I staggered outside. Then I threw up, the vomit stinging as it came out of my mouth and nostrils, and I felt light-headed.

The next thing I was aware of was Gordon putting his hand on my shoulder and asking me if I was all right. 'Why, I'm only having a pee,' I said. 'Who were you talking to then?' he asked. 'You've been out here for five minutes.' The mist in my mind cleared for just a second. I thought I remembered it – I had been having a conversation with the large boulder at my feet. It was like being woken suddenly from a dream, and seconds later could not remember what I, or the rock, had said. Bizarre.

My trusted assistant Gordon seemed to enjoy my predicament. But he wasn't yet aware of his own situation. He took it like a trooper, then ran outside to be sick, too. The sound of him throwing up made Julio and the *shaman* chuckle. No danger of us ever becoming addicted *yoppo* sniffers! Gordon appeared through the tiny entrance, wiping the dark dribbles from his chin, and Julio moved to leave the long-house. Gordon pointed to him and mimed *yoppo*-taking to Panutsa. He got the meaning straight away and called Julio back. It was Gordon and I who sniggered now. 'Sheet guys, that was worse than kissin my murther-in-law,' Julio said, rubbing his watering eyes afterwards.

The next part of the ritual was to follow when the *shaman* had prepared himself in private for the next stage of the ceremony. Whatever that involved, we would never know. I moved outside, and sat with the two hunters, Lanyetsa and Wariena. I could hear the snorting and gagging sounds continue from inside the nearby long-house. The other two men were readying their weapons by a small fire. Two blowpipes were propped against the side of the thatched roof, each over nine feet in length. The men were squatting by the glowing embers making the blowdarts and tipping them with curare, a deadly muscle relaxant that the Indians extract from a forest vine. They paint the poison onto the darts which are then dried over the fire.

It was half an hour before the *shaman* appeared, wearing the tarantula head mask. I wondered how he could even walk after that *yoppo*-sniffing session. He joined the two men and began to chant, all the time beating his chest lightly with one clenched hand. The lyrics continued for ten minutes. He was asking his spirits for good fortune during the hunt. Once the darts were prepared, we hoisted the rucksacks with our equipment onto our backs and were on our way into the forest, to watch these people hunt for food as they have always done. By now it was well past midnight and the forest was dense and black.

As we trekked off, I could not help thinking about our own lost culture, and whether just the fact of our being there was the thin end of the wedge for these friendly people. I learnt much later that the process had begun many years earlier and that these Indians were facing changes forced upon them by successive national authorities. Charlie Brewer told me a story about a visit he made with a working party of government people from Caracas. One official introduced himself as having come from the country's capital, Caracas. The headman's first question was, 'What is Caracas?' To them the forest was the world.

As we set off on foot, weighed down by our packs of equipment, we were soon hit by the warmth and humidity of the forest. We were dripping in perspiration, trying to keep up with the hunters who moved forward with such ease and lightness of foot. For two dark hours we stumbled on through the tangled undergrowth, closing on the summit of the brave mountain, Punto Bravo, a gigantic rock formation poking through the canopy of this rich forest.

It was still night as we struggled up the steep black granite slopes, slipping over tree roots and stones. The Piaroa must have thought us such cumbersome creatures. Eventually we drew level with the canopy of the surrounding forest. I set the camera up, with dawn only minutes away. We had only just made it. Julio lay on the soaking wet granite rock gasping for breath.

But there was no time for us to sit and recover, Gordon had to set up the sound recording gear, disappearing behind scrub bushes 300 feet away, while

I loaded film. With the camera set up, I surveyed the scene while trying catch my own breath. It was getting lighter every second as the sun neared the moment it would show itself to us. What a *view*, what *sounds*. A new day was coming and I hoped Gordon would be getting it all in glorious stereo.

A pair of scarlet macaws flew by, calling raucously, and settled into a tree to our north. Panutsa stood erect watching his forest and, right on cue, the orb of fire came over the ridge of the distant mountains backlighting his primeval-looking form, the blowpipe over his shoulder. There was no doubt in my mind that he was enjoying the view, too; there was even a slight smile on his face. The camera rolled and trapped within its mechanical housing a moment in time no one will ever see again in quite the same way. The cloud formations, the *shaman*, the light, were all unique for us alone in that wilderness.

The hunters were anxious to get going although they did allow me to indulge in the view, watching the ever-changing scene as the sun rose higher. I was the one who eventually gave in as the swarms of tiny blackfly cloaked me. Each breath drew in one or two up my nostrils or into my mouth. They were inescapable. I went over to where Gordon had hidden himself to get a sound recording of the atmosphere and he awoke with a start when I called him. 'Yeah, great wasn't it?' But what I didn't know until much later was that he had fallen asleep and thought he might have spoiled it all by snoring so close to the microphone! We had a good laugh about it – but only after we had listened to the perfectly recorded dawn chorus.

The Piaroa dropped into the darkness of the forest once more. They moved stealthily ahead of us, blowpipes in hand, eyes scanning for the slightest movement, ears finely tuned to every sound. Then Gordon farted loudly and the effect on the Piaroa was hilarious – they literally had trouble standing they laughed so much. They all cheered and chorused the words '*kariminay dakwayo sofu*'. Gordon asked Julio to find out what they had shouted and the *shaman* told him it meant 'white man's bottom is making noises and is about to shit'. We laughed until we cried. Infantile humour, I know!

The Piaroa's skill in the use of the blowpipes became evident. They could catch a bird from a treetop 100 to 130 feet away. They did miss, but not often. Sometimes they would have to climb high trees to retrieve the snagged quarry from the branches. They were obviously accomplished climbers. They almost ran up the tree trunks, or so it us. Small quarry, like birds and rodents, would be tied with hung about six feet off the ground to be collected later on the to the village.

the men stay out overnight in the forest, they construct a temporary shelter from palm and wild banana plant leaves, and fashion hammocks from vines. All this is done in a matter of half an hour. They rarely keep a fire going at night after they have eaten, unless they have seen a jaguar, or its paw prints, close by during the day's hunt. The jaguar is the one animal they fear above all others.

It was on the second day of hunting that the men pointed out some holes in the leaf litter on the forest floor. Some were obviously excavated rodent or small mammal dens. Wariena snapped off a piece of vine and crouched at the entrance to one of the burrows. I noticed some fine strands of silk in the entrance, caught by a fleck of the new day's sunlight breaking through the overhead vegetation. With his left hand, he cautiously poked the vine into the opening and began to twiddle it between thumb and forefinger. He muttered something to the *shaman* and began to pull the vine out very slowly.

I saw her front feet first, padding delicately but urgently after this intruder. She came out in the dappled sunlight in front of Wariena. She was a splendidly large tarantula and was not going to be distracted from her chase. He lifted his right hand, not moving the vine in his left, moved his arm over her body and lowered his hand to just above the spider's thorax. Then he made a quick movement, thumb first, pinning her to the leaf litter. With great care her eight legs were folded back behind her abdomen. She was caught and helpless. If a sorrowful tone comes through these words it is exactly as I felt, though this emotion was tempered by the reality that these men needed food to survive – no quick trips to the supermarket for them – or at least regarded the giant tarantula as a prized delicacy, rather like we feel about lobsters or caviar.

Wariena pulled a broad leaf from its stalk and expertly wrapped the tarantula with it, parcelling it with a strand of vine. She was trussed alive, unharmed, and would remain so until the hunters became hungry. The spiders are kept alive because, if killed, the meat would go off quickly in the high temperature and humidity. In less than 20 minutes they caught another four spiders.

We had been travelling for seven hours and were all hungry. We had only stopped briefly to drink the crystal-clear cold water when crossing creeks. The hunters had pulled leaves from a nearby plant and curled them to fashion drinking cups. We made pathetic attempts to copy them and they laughed before making cups for us.

We were now looking up through a break in the canopy at a large, black rock formation. A huge boulder curved downwards and where it met the forest floor it formed fissures and caves. The wall of the rock was painted with ancient shapes – we discovered much later that they were about 3,000 years old. It was a dark atmospheric place. I sensed, too, that this was no ordinary hunting pit stop. It was overpowering within its strangeness, although it did not create a sense of fear.

Gordon and I were transfixed, trying to absorb it all. On the rock's side I could clearly make out a sun, a lizard, and then – so relevant to our story – a hunting scene showing a deer kill. My eyes followed the rock down to the base where the medicine man stood with his blowpipe over his shoulder. I tried to picture the Indians painting here thousands of years ago. Nothing here has changed in all that time. Or has it? We were there and I had a feeling of guilt. My sense of intrusion grew when we saw that the slits at the rock's base were funeral holes, the bones of the Piaroa people buried there in front of us. The heat and the strangeness were overwhelming.

This was all of no matter to the Indians – they were simply hungry. The three of them gathered by a small depression in the ground to prepare a fire. They showed us how to make fire by rubbing two pieces of wood together. It took a good five minutes of hard wood spinning between their palms to produce a handful of wood shavings that Lanyetsa cupped in his hands and gently blew upon. A glow emanated from the centre of the furry ball. For me, looking back, it was the most poignant moment of the entire trip. *Homo sapiens* making fire as he first did, how long ago? The glow in his cupped hands was somehow fittingly symbolic.

My thoughts were overtaken by the need to film all of this. I felt bad asking them to make another fire just for the camera, but that's the filming business. Although they may not have understood what we were doing or why, they didn't mind, in fact they had a good laugh at our expense most of the time as we sweated and toiled with our cameras. Having seen the film footage, it is wonderful but, however good, film emulsion can never capture the real moment that one is absorbed in by just by being there. Temperatures and smells make an enormous difference to the scene.

With the fire underway, the spiders were unwrapped and instantly dispatched with a wooden spike pushed through the cephalothorax, the head and chest segment. Lunch was about to be served. The two formidable fangs were removed and kept to one side, the large abdomen was twisted off and the contents squeezed out onto a leaf. A large blob of grey and yellow slime oozed out, containing about 40 eggs. The leaf was then wrapped parcel-fashion and placed in the hot ashes at the edge of the fire for a couple of minutes. When cooked a miniature omelette was the effect. I found the taste bitter and horrible, but I was extremely hungry.

The rest of the spider was put over the flames to roast for a few minutes and then was eaten in exactly the same way as a crab. This was delicious. It was fiddly to get the meat out but it was crab-like in texture and, I thought, tasty. The hunters ate theirs quickly but we approached ours a little more timidly. The *shaman* then picked up one of the tarantula's fangs that had been put to one side and used it as a toothpick! Of course we had to do the same, so I picked away with a fang, too, and I have kept it, along with the two pieces of wood that they used to make the fire, as a special keepsake.

We had been filming and taking part in all of this while staying close to the fire. But we still needed to interview the men about their sacred mountain and why they eat the tarantulas. To do this, we moved down the curved surface rock to the forest floor – scattering several basking lizards on our way – to find a more convenient place to set up our camera and sound recorder. The smoke of our fire had been a blessing because it had kept away the flies, but at the base of the rocks they were suddenly worse than back in the village. It was almost impossible to work there and Gordon had to set up the recorder on remote to allow him to waft the flies away from me while I filmed. It was the most uncomfortable place I have ever filmed at. Leaving, however, was a luxury we could not contemplate until we had shot the interviews.

After Panutsa had told us about the sacred mountain, Julio asked Panutsa why the Piaroa ate the spiders. Gordon was wafting away behind me telling me how many of the flies were on my back while I was trying to think about the deep spiritual meaning of dining on tarantulas. The *shaman*'s reply, 'because they are delicious,' made Julio laugh so much that it ruined the recording. We then had to do a second take and suffer the flies for another 15 minutes. The *shaman* told us as we left that they call them body flies. It didn't take too much thought before we realised why.

So we left the funeral caves with good film and interviews, knowing we had been breathing in flies that were helping to decompose bodies. Thankfully at that time we were not aware that we would have to return there two months later to do some retakes after a local airline had lost one of the cans of film. It is a cameraman's worst nightmare and the first time that it had happened to me. Of course it had to be *that* roll which was lost. For once in my life, I can thank an insurance company because the original, as we call the negative film, was insured against this kind of loss. Les Marshall, my broker, pulled out all the stops with no quibbles to get us back to re-shoot the lost film. That is how we ended up returning to the site we nicknamed 'hell on earth'.

We knew our return was the final part of the shoot and, just for this occasion, I had stashed a bottle of champagne in one of the cases for an end-of-shoot celebration. It is a tradition in the film industry to have what is called a 'wrap party', film-speak for when a production is finished or wrapped up. This time we squatted around the fire simply to keep the flies away and, while the Piaroa roasted giant tarantulas, I opened the warm bottle of Moët et Chandon and handed around small plastic cups of the stuff. The three men sniffed the champagne, their eyes squinting as the bubbles fizzed into their faces, before putting the cups to their lips. Wariena smiled at his partners and they began giggling like schoolboys. I suppose it was just another odd thing that these foreigners did as far as they were concerned. Sitting in those funeral caves, munching on spiders and swilling them down with warm bubbly must have been the strangest wrap party of all time. The Piaroa men loved the champagne, but had the truth been told, Gordon and I would have preferred a pint of cold lager to wash our tarantulas down.

Our time with the Piaroa Indians came to an end all too quickly. It had been such a special few weeks for us, and I felt genuinely sad at having to say goodbye. I was particularly sad to leave our three hunters, Lanyetsa, Wariena and Panutsa. We had shared their lives and, in a different way, they ours. One morning they had even taken the trouble to teach us how to use their blowpipes. They set some papaya fruit up on a raised piece of ground by one of the huts and showed us how to hold the pipe. A blowpipe is about ten feet long and fashioned from three separate pieces of wood. The outer tube is made from a hardwood, painstakingly selected for its straightness, and then whittled away at the inside for days to hollow it out. After this a length of reed piping that acts as the barrel is inserted and, finally, the Piaroa fit a carved mouth-part like a large trumpet mouthpiece to one end. This is glued in place with wild beeswax. Sometimes they also use wild peccary or paca teeth, or seeds, as sights which help them to aim. They set these into the length of the blowpipe about 18 inches from the end.

The first thing that struck both Gordon and me was the small amount of puff needed to send a blowdart at incredible speed such a long way. The most difficult part to master was holding the pipe, which you had to do with both hands, and sighting it accurately. Practice makes perfect though, and it was only an hour before we were both hitting a papaya fruit from 30 to 50 feet. The men never laughed at our efforts, but they obviously enjoyed the spectacle of us learning. Although it is doubtful that our early promise with the blowpipe will ever be of use at home, it is one of the memories I shall cherish.

As we were packing our equipment for the following morning's journey back to Ayacucho, Julio, Gordon and myself opened three cans of Venezuelan Polar beer. We were settling down in our hut for the last night, listening to Julio's terrible jokes, when the *shaman*'s head appeared through the hut's doorway. 'Pssst,' he said. Gordon made a puerile joke 'not yet but we hope to be later.' 'Pssst,' he said again, flicking his thumb upright as he said it. I thought he was trying to attract our attention but Julio pointed out that he was mimicking the sound the beer cans made when we opened them. He simply wanted to drink one with us. We invited him into our hut and had several more cans together, finally falling asleep satisfied and happy. Panutsa liked the beer 'too much', Julio told me.

Chapter Six

Love on a Film Set and the Killer Instinct

One aspect of the giant tarantula that I desperately wanted to film was its reproductive behaviour. I knew there was no chance of capturing this on film in the true wild so we decided to build an artificial set. To have been able to film one in its natural setting, at night, we would have had to have used a generator and the noise from the motor, together with the bright lights, would have sent the tarantula scuttling into its burrow. Hence the artificial set. But first we needed some spiders.

Rick had been running about like a madman visiting Piaroa communities closer to Ayacucho. Against all odds, he finally persuaded some Piaroa to show him where they caught their spiders. The idea was that they might also give him a few for our close-up set filming. These people had once lived in the forest, but now existed on the outskirts of town in government housing projects. However, Rick's first outing, with Jorge and Julio, didn't quite go according to plan. This is an extract from Rick's diary:

> The Land Rover decelerated and came to a stop. I snapped back to reality. Julio quietly said, 'Get out, we go into the forest to see some Piaroa here.' Here! I couldn't even see a trail. I thought, 'Yeah right, they're going to take me on a never-ending search for the elusive giant tarantula.' To my surprise, Jorge and Julio walked about 40 feet up the road then cut straight into a wall of vegetation. While in nimble pursuit, it wasn't until inside the forest that I could make out a small man-made trail. As I strained to keep up with my guides, I tried to keep one eye on the flora and fauna and another open for snakes. The forest was extremely humid and dark. The musty smell of decomposing leaf litter filled the air. The silence was broken by the sounds of buzzing cicadas and bird calls which filtered down from the luminescent tree canopy. It wasn't long before my clothes were soaked with sweat and clung to me like a shabby skin. Much to Jorge's and Julio's amusement, I'd veer left or right off the trail to explore small tarantula-like burrows or interesting plants, frogs and insects.

About half a mile into the forest and on one of the many streams which criss-crossed our trail, we came upon a clearing. As we entered what was obviously a man-made garden our approach was announced by barking dogs and the sounds of Piaroa children playing in the stream. This Piaroa community turned out to be one large thatched communal hut in a small clearing devoid of vegetation. Although highly interesting, I thought to myself, 'Is this it?' A young Piaroa man, wearing a loin cloth and carrying a machete, approached and spoke to Jorge. I was gestured to come forward by Jorge and I think I was introduced. I smiled and nodded. For all I know I was being announced as a large, fat, rich, crazy, white-eyed devil trying to buy tarantulas. As the two men talked, I gazed around the compound taking in the natural serenity of it. I watched several little children playing in a section of stream that had been dammed to make a pool; they seemed oblivious to the perils of large or venomous snakes which I thought lurked under every log or pile of leaves. Next to the hut a young, heavy-breasted woman sat nonchalantly nursing a new-born child. Save for a scanty red cloth covering her genitals she was nude. I tried not to stare and, as a Westerner, felt some unexplainable guilt for looking upon this woman and invading this tranquil family gathering. However, they made me feel at ease.

Julio and Jorge excitedly spouted that the Piaroa man's old mother had caught some monkey spiders for me the previous night. As everyone manoeuvred me in front of the hut, like a kid at a surprise party, the old Piaroa woman emerged through the thatched entrance and straightened up. There in front of me stood a four-foot sinewy, naked, toothless, grinning old woman holding up six live giant tarantulas individually packaged in palm leaves… I was aghast at both sights. Through Julio, I asked her what she wanted for them – thinking money, lighters, fish hooks, and so on. I immediately found out how shrewd she was – she wanted a brand new 'large' aluminium cookpot and wasn't going to settle for anything less. Needless to say I was fresh out of large new cookpots. With a promise to return tomorrow I reluctantly had to leave the tarantulas behind. Even Jorge's charm wouldn't change her mind. As we wearily trudged back to the Land Rover, I asked Julio what might become of the spiders if we couldn't return tomorrow. 'Haay buddy, they'll get eaten!' he said in his unique style.

As I sat in the vehicle, which proceeded on to the next community, I could only guess at what other disappointments awaited me and what I'd tell Nick when I got back. If I told him I had seen the giant tarantulas but had had to leave them behind, he'll blow a gasket! After some distance further down the dusty dirt road we rounded a bend to where the forest opened up to another Piaroa village. This one was much larger, and at the end of the road. Tall dark forest encircled it on three sides. Again my hopes sank as the first large thatched hut I saw had a cross on it; it was a church. Beyond the church was a stone-surfaced street about 100 yards long which ended at a single street lamp.

As we climbed down from the Land Rover we were literally mobbed by children and adults, as though we were rock stars at the stage door of some theatre. It took me a few minutes to realise that they were gradually pushing us into the centre of the street waving lots of objects in the air. There must have been more than 50 people. Thank God I am not claustrophobic – or afraid of spiders!

It became apparent that Jorge had exaggerated my needs to the Piaroa. Every person present, it seemed, was waving at least three to six live giant tarantulas at me. It is a sight I shall never forget. I told Jorge in no uncertain terms that I had said I only wanted ten specimens at the most. 'No worry, no worry,' he shouted as he waded into the crowd as mediator and 'giant tarantula picker'. As I tried to look for the largest spiders, I was beginning to prickle all over my arms and face. As the intensity of the irritation grew, I suddenly realised it was the urticating hairs being shaken from the tarantulas as people waved them about wildly over my head.

The heat and humidity being given off by all the crowding sweaty people, combined with the severe itching from the tarantula hairs, was too much for me to take. I hurriedly gathered the packets of spiders from Jorge and made a dash for the vehicle. It was like breaking the water's surface to gasp for fresh cool air...

Rick's wanderings finally paid off, and he did return the following day with a brand new 'large' cookpot for the old Piaroa woman. The most worrying fact that emerged from his encounters was that none of the resettled Piaroa villages used the tarantulas in any spiritual ceremonies. I began to wonder whether we were on a 'wild tarantula chase'.

It took Gordon a week to construct our film set. The masterpiece was to be complete with subterranean tarantula burrow and above it a forest floor with living plants and leaf litter. Gathering the materials from the forest floor five feet below was not a problem but, as Rick pointed out, we then not only had to make an authentic looking set to film, but more importantly, had to introduce a female spider, accustom her to the lights and convince her that it was home.

The set measured ten feet long by six feet wide, and looked like a huge chest-height table with a tent darkroom at one end of it. Gordon used chicken wire, paper, wood and mud to build it. The underground chamber had 'filming windows' – two ports cut into it with removable glass panels so I could film anything that went on inside. We had carried these panels all the way to Venezuela from Blackpool in Lancashire (where friends of mine, Chris and Davia Walmsley, run a glass design company). They had laughed hysterically when I had asked them to provide the panels and explained what they were for.

We used specialised cold fibre optic light sources to illuminate the interior of the burrow because any strong heat would have killed the tarantula. At the end of the first week Rick showed our spider around her new home. She scuttled in and headed for the farthest corner. Then we had a further week of worry until one morning we discovered a newly-laid bed of silk on the burrow's floor. 'She's decided it's home, guy,' Rick said. She had finally settled in and regarded it as her territory. We could begin to film at last.

At night she would move to the entrance and wait for passing prey – perfectly normal behaviour for a tarantula. We did not have to help her too much with catering facilities as all manner of insects and frogs would visit the set from the real forest floor, just below. Frogs and beetles were the most common, and moths, too, as well as something we never expected – snakes.

The chamber and tunnel were completely covered by my filming hide, with yards of black-out material festooning the whole contraption so that the tarantula could not see any light or frightening movements from me when she was in the burrow. I used to check the set regularly and one morning raised the entrance flap to the hide, ducked in, only to be faced with a six-foot long serpent with a frog protruding from its mouth. The snake was a kind of racer, and the thickness of an expensive cigar. Rick told me it may be poisonous, but that the fangs were at the back of the mouth and so not really dangerous to humans, 'unless you stick your finger down its throat, guy'. Despite his reassurance, I hurriedly closed the hide entrance and let it continue its meal. I never entered the hide so casually again.

As almost nothing was known about *Theraphosa blondi*, we had to work with assumptions based on other species of tarantulas. Having Rick on the set was invaluable as he could always tell me what particular behaviour meant when it happened in front of the camera. For example, there was a period of three nights when our spider wouldn't come out from the burrow tunnel, preferring the shade. Rick thought the film lights we were using above ground to illuminate the floor area were too warm for her, and so we moved them a little further away. Then she emerged almost immediately.

Our female had well and truly settled in when we decided to introduce a suitor to her. We knew that the tarantula males were preoccupied with copulation; that they wandered about the forest floor looking for females and, if not killed by them during the coupling, would die in any case soon afterwards. Their lifespans were so much shorter than the females'. Rick had searched the forests close to town for a male tarantula and, although we saw dozens of females, he only found two males. None of us could understand this.

Before the sexually mature male sets off to find a receptive female he sheds his outer skeleton. Several days later, once his new covering has hardened, he constructs a 'sperm web' between two objects such as tree roots or rocks. The spider crawls upside down along the web and makes a fine silk canopy by swaying his spinnerets, the finger-like organs on the rear of the abdomen where the silk comes out, from side to side. He then deposits droplets of sperm on the underside of the web. After crawling out, he carefully moves across the top of the silk and gently strums his palps, the two shorter leg-like limbs at the front, over and under the edge of the silk canopy, dabbling the little sacs on his palps' under-tips into the sperm droplets. By means of capillary action, the sperm is drawn up into the sacs and, this completed, he begins his search for a female.

It was midnight, and our forest floor was lit by artificial film lights powered by a portable generator. The female had moved to her burrow entrance to wait for supper and we knew the heat from our lights would soon send her scuttling back into the cool of her den. We did not have much time.

Rick placed the male in dense vegetation to see if his natural instincts would take over. After ten minutes he began to move out of the undergrowth onto the leaf litter in front of the female. She twitched, obviously sensing a presence. Her delicate feet pads, I thought, would be telling her that dinner may be approaching. Rick reminded us afterwards that she is equipped with highly sensitive chemoreceptors and would have known it was a male spider. Well, she might have done if there hadn't been a power cut at that very moment and our generator suddenly broke down. It was all over for the night as she darted back into the burrow while we fumbled for a torch. Hours of preparation for nothing.

Our male could so easily have ended up as a meal for her. It's well known that if a female tarantula is not receptive and a pushy male perseveres with unwanted advances, he may be pounced upon and even devoured. The following night we set everything up again, hoping that the newly serviced generator would hold out. The male was placed in the set again, having spent the previous night in our room, and told by Rick to 'go break a leg'. He started to make slow progress towards her burrow – he knew exactly what he was after.

The coming together was not exactly a scene from *Wuthering Heights*, but it was fascinating to witness. His front legs tapped the ground and the female's front legs. Her fangs protruded menacingly as he began to push her backwards and upwards at the same time. She towered over him, her fangs just above his head. At the extremity of his two palps were the little sacs filled with seminal fluid, and he tried to hook one of them into her epigenum, the receptacle fold or reproductive opening on the underside of her abdomen. She was not having any of it. Sensing this, his reaction was incredibly swift, almost too fast for the human eye to register. In a fraction of a second he was 18 inches away – his life intact. We breathed a sigh of relief for him.

At this stage in human affairs I would have sloped off home. Not a tarantula, though. He advanced again and I was convinced he was pushing his luck too far. It would be curtains for him this time. In the event it was not and he successfully inseminated her with both sperm sacs. The mating was in part a kind of dance with him tapping her and the ground until he was in the correct position. It was strangely touching to be close to all of this and we were only brought back down to earth by Gordon leaning across the film set and saying to the female, as she retreated to the cooler shade of her burrow, 'How was it for you?'

Rick could hardly contain his excitement. 'Do you realise that's the first time that has ever been seen, never mind filmed?' The three of us stood there celebrating, with mineral water, taking in the amazing scene we had captured on film. Things were really beginning to come together at last.

When we weren't filming our tarantulas, we kept them in large Tupperware boxes in our accommodation. We kept several because tarantulas become easily stressed and we wanted to make sure we didn't tire any one individual out. Rick would always tell us when we should 'bring on the stand-in'.

Our initial wariness subsided and we soon found ourselves dealing with the tarantulas quite casually on a day-to-day basis, cleaning their quarters and feeding them. There were times when Gordon washed out their Tupperware boxes, and the hairs the spiders released caused his chest, arms and neck to become so painful that he had to spend hours under the shower to get relief from the burning sensation. On several occasions one of us would be woken at night by a noise, and in the beam of a torch-light discover an eight-inch tarantula climbing the bedpost or coming out of one of our suitcases. We would scoop it back into its box with the lid – and one day even learnt to tape the boxes shut! We had not come across anyone who had been bitten by *Theraphosa blondi*, although we noticed the Indians handled them with great care. But that was all before the *Bothrops atrox* sequence was filmed. Better known as the *fer de lance*, it is one of the most poisonous serpents in South America and was to change our perception of the spider for good.

I was filming a female *Theraphosa* in her burrow on the set when Rick shouted that a snake was close. He quickly identified it as a *fer de lance.* Snakes wander the forest floor day and night in search of prey and frequently investigate rodent or mammal burrows for food or rest. I did not want our female tarantula to end up as a snake's snack but I could not intervene.

The snake entered the burrow tunnel and I watched through my camera as the tarantula immediately detected its vibrations. She spun around to face the opening where the tunnel widened out to form her chamber. I saw the snake's head coming into the camera's frame and could feel my heart pounding. The tarantula made a lightning dash forwards, the snake momentarily coiled over her before it continued moving forward, its lithe body slithering past the spider who turned to face the snake inside the chamber. The *fer de lance* took up a strike posture. The spider looked quite normal. Had she been bitten?

I shouted to Rick and Gordon who were outside the hide telling them what had happened, but before anyone could reply the snake yawned and fell limp on the earth floor of the burrow. It was dead. The tarantula had bitten the *fer de lance* and it had taken less that two minutes for its venom to kill it.

Still bubbling with excitement over this turn of events, I continued to film as the spider moved towards the dead snake and picked it up by the head with her fangs. During the whole episode I hadn't consciously realised I was only inches away from such a deadly snake. I was so consumed by the fascinating behaviour and the luck of having captured it on film. Now I was already thinking of it as an opening sequence to grab the audience's attention and, hopefully, prevent them changing channels for the following 54 minutes.

I was still inside the film hide 15 hours later. Hunger had been forgotten although the crate of diet Pepsi and mineral water that Julio had supplied the previous day was almost empty. I had lost all sense of time and was finding it difficult to keep myself awake. I shook my head from side to side, slapped myself on the back of my neck and tried to convince an exhausted body that it wasn't ready for sleep just yet.

It took all of those 15 hours for the spider to devour every bit of flesh and bone of the two-foot snake, save only for the skin. I was amazed at her capacity because the only visible change in her size was now a bulging abdomen. She looked ready to burst.

As befits a good housekeeper, the tarantula removed the snake's skin from her home, depositing it on the leaf litter outside for the ants to finish off. This was not as altruistic as it sounds – Rick explained that it is vital for her to remove any animal remains as they attract a parasitic fly that lays its eggs on the food as well as on the spider's abdomen. The fly's eggs develop externally but when the maggots hatch they burrow into the tarantula's flesh and consume it alive.

Unfortunately she didn't carry the skin out immediately she had finished eating, and I must have fallen asleep because I woke up with a start to find the burrow empty.

Horrified at having missed this final bit of behaviour I lifted the flap of the film hide and crawled out, my legs aching from being in one position for so long, to find Gordon asleep on a log, snoring soundly. Our vigil had come to an end, almost. I looked into the film set and saw her moving across the leaf litter. She had dropped the snake's skin behind her. I don't give up that easily, so I gingerly put my hand in and picked up the remains. Unlike Rick, the thought of such a big spider walking over me makes my hair stand on end.

Watching her go back into the tunnel was the cue to hurry back into the film hide, replace the skin on the floor of her burrow and put the filming window back in place. Ten minutes later she returned to her lair, picked up the snake skin with her fangs and carried it out again. This time the camera was running. Instinct is a wonderful thing, be it the snake's or mine.

Another formidable snake we filmed was the anaconda, arguably the largest snake in the world. This incredible aquatic serpent can grow to well in excess of 30 feet. We were in a small dark creek, a backwater of the Orinoco, filming a female giant tarantula drinking at the water's edge. Something caught Rick's attention on the other bank. He had seen a beautifully coloured anaconda resting on a fallen log in the water. It was a fairly young one, about 10 feet long, and let me come to within inches of its head for some shots. Its forked tongue flicked out tasting the air.

Although the anaconda is not venomous, it does have a very powerful bite and it kills its prey by constriction, squeezing the lungs until they cannot breathe any longer. I noticed many parasitic tics on the upper half of its strong body. They must have been picked up from the leaf litter as it travelled the land alongside the local creeks. After ten or so minutes the anaconda moved slowly into the water, the dark tannin-stained depths of the creek swallowing her image.

The way she consumed the snake represents only one way that *blondi* deals with its prey. With some kills it uses another amazing device, its spinnerets.

These are two one-inch long finger-like protrusions on the rear of her abdomen. When the tarantula has killed its victim it uses its fangs to inject large quantities of venom. This acts like acid on the tissue, turning it into a mush that the spider sucks in. Her fangs are also used as utensils, which she can move independently to draw the melting mush into her mouth. As the body of the prey slowly disintegrates, it falls apart. With smaller prey, she shrouds her victim in silk. The tarantula first lays a bed of silk from the spinnerets. She then places the prey on the silk mat and covers it with more layers of the finely spun gossamer. It takes a long time to do this – half an hour with a grasshopper, for example – and once her meal is wrapped in this way she picks it up and commences eating or sucking. This remarkable behaviour is fascinating to watch at close quarters but incredibly ghoulish. I have long been convinced that horror film makers get their ideas from natural history.

Imagine a wasp as big as your hand with a sting that makes a hornet tame by comparison. The stuff of your worst nightmares. It does exist in the rainforests of Amazonia. Its local name is the spider wasp and its scientific genus is *Pepsis.* It feeds on the nectar of flowers and, aside from man, is the only other predator of the largest spider in the world. This wasp scours the banks of the rivers and creeks in search of *Theraphosa blondi* and, as befits a flying demon of such proportions, makes quite a noise as it flies low and fast seeking its quarry.

It alights on the leaf litter near a burrow and darts head down, zigzagging to and fro until it picks up a target. *Theraphosa* the hunter suddenly becomes the hunted. The wasp grapples with the tarantula, usually outside the spider's burrow, though it is not averse to diving into the burrow if the spider is below the ground. It will then entice the spider out into the open, like the Piaroa do with vine stems, by mimicking the tapping movements of the male spider. The vibrations bring her out, her thoughts on mating.

Once locked in combat, the wasp flips itself onto its back and moves underneath the spider where it seeks out a soft spot on the abdomen to deliver [illegible] paralysing sting. If successful, the tarantula's nervous system is [illegible]d and it is unable to move. This soporific state can apparently last for many weeks. The wasp then drags its sleeping partner into the burrow and lays its egg on the tarantula's body. Finally the wasp seals up the tarantula's burrow, and flies away. Over the following weeks, the wasp's egg develops on a live but immobile larder of fresh food. The parasitic grubs eat away until the spider dies, and the wasp is then ready to emerge from the tarantula's tomb as an adult, digging its way out after pupating.

We witnessed two variations on the wasp-tarantula battles. The first was when a spider wasp attacked but was repelled three times by a female *blondi*, the wasp eventually leaving the scene in a groggy state. The second was when a *Pepsis* wasp attacked a tarantula but then itself became the victim, having received a lethal bite.

Wasps do not always win, it seems; tarantulas have their own sophisticated defence mechanism, whether a venomous bite or their hairs. The spider produces a snake-like hiss by rubbing together stout hairs and striker spines located between the leg segments of the first two pairs of legs and the palps. This warning acts as a deterrent to hopeful predators. The stridulation is sometimes done alone and on other occasions in unison with sending urticating hairs into the air.

The release of hair is done by using spines that protrude from the inner surfaces of the two hind legs, and these can be clearly seen by the naked eye. They differ substantially from the long hairs that cover the same limbs and give the spider a furry look. When the spider is threatened, it uses them to considerable effect. On one occasion Charlie removed three white mice that he had put into the tarantula vivariums for food but which had not been eaten for some reason. They were still alive, but within an hour all had died. They had inhaled the hairs that the spiders had rubbed off their abdomens.

The first time I saw the use of such effective defence behaviour in the wild was while filming a female giant tarantula moving across the forest floor. We heard something cause a rustle in the undergrowth, and withdrew with our cameras to a quiet spot by the base of a tree. It was some minutes before we saw the coatimundi rooting for grubs and insects with its long peculiar nose. The size of a small dog, the coatimundi is related to the racoon family. It has a long, banded tail nearly as long as its body, short ears and is a reddish brown colour. The most remarkable characteristic for me is its snout. It is almost a fifth limb in the way it can bend backwards, forwards and sideways.

This coatimundi came across the tarantula and immediately launched an attack, clawing and padding at the spider with both its front feet. The tarantula's response was instantaneous. The urticating hairs floated up, the sun back-lighting them and allowing us to see the fine microscopic grains flickering like dust. The coati's amazing nose began to pour with fluid and within half a minute it retreated, rubbing its nostrils with both paws.

The coati eventually continued its search for more accessible prey, getting relief from pushing its nose into the soil and leaf litter. The spider, which had been tumbled over and over, lay still on her back but after a couple of minutes righted herself and continued her journey to the safety of a nearby burrow.

Our next stage was to concentrate on finding and filming other species of tarantulas. One evening, armed with our heavy-duty Maglite torches, camera bags, and Rick's bug-collecting contraptions, we walked along a trail into the forest. Julio tramped along carrying a six-foot wooden ladder. The forest at night is a very different place from the forest of daylight hours. Not least the sounds. Instead of ant birds, cicadas and chattering monkeys, there are nocturnal creatures whose calls, at times, stop you in your tracks.

Our timing was perfect. It was almost a full moon, and a common potoo was calling. Rick caught it perched on a high branch in the beam of his torch light. It is a most peculiar bird to look at, resembling a stump of old wood but it also has the most touching call of any bird I have ever heard. It cries out four mournful descending notes that carry far in the still of the night. Its song can only be described as making it sound lonely. It seems to wail 'poor me all alone'. In some places it is called just that – the 'poor me all alone' bird.

Further along the trail, Rick suddenly called out, 'Here we are, guy, what do you think of them?' Three feet above our heads was a spray of exquisite pale purple orchids. Five flowers, each one the size of a saucer. I simply couldn't have imagined such delicate hues existed in the forest at night.

'*Psalmopoeus irminia* lives in there,' Rick said.

'Oh really, you mean in those *Catleya superba*,' I replied cockily.

'That's right, guy,' he added smiling.

'OK, Rick, what is the salmonella hernia or whatever you called it?' He laughed and prodded behind two of the flowers. Suddenly a black furry thing jumped out onto the nearest branch.

She was velvet black, extremely hairy and her legs looked feathery. Each leg had a fine golden stripe along its length. I just had to get a photograph of it among those beautiful orchids. Julio put the ladder against the tree trunk and Gordon steadied it as I climbed up to the level of the spider. Rick was chatting to me as I fiddled with flash settings, 'It has heavily scopulated...'

'For God's sake, Rick,' I butted in, 'speak English will you? What the hell is scopulated?'

'Padded, you turkey, and those pads and their legs act as air foils if they jump off a branch and fall to the ground or into water. They can use them like paddles.'

At that moment, as though Rick had set the whole thing up as a prank, the five-inch spider leapt onto me and, within a fraction of a heartbeat, it darted halfway down my right arm and then reversed up under my shirt sleeve, settling somewhere close to my right nipple! I froze with utter panic. I couldn't even find the courage to shout at Gordon and Julio to stop laughing – because that would have meant me having to *breathe*, and that was definitely out of the question.

Rigid with fear, my descent of the ladder was the nearest thing to reverse levitation by some unseen force you could *ever* have seen. I did not want to add this tarantula to the list of creatures that I had been bitten by. Rick sauntered over to me at the foot of the ladder as though we had all the time in the world. He lifted my shirt and, unbelievably, looked closely at it, trying to sex it.

'Get it off,' I whispered like a ventriloquist. He manoeuvred a collecting jar towards me and, in the blink of an eye, I was liberated. The tarantula jumped onto Rick instead, scuttling into his shirt!

'Haay, the thing like hairy chests,' Julio laughed. I prayed for it to leap onto Julio, but it didn't. Rick, cool as ever, slowly peeled his shirt off and persuaded it into his collecting pot with no further trouble.

I moved the ladder beyond *Psalmopoeus*-leaping distance and put a telephoto lens on my camera. We spent a pleasant half-hour taking pictures of it before leaving her in peace among the orchids.

With my pulse rate returning to normal, we trudged off to another site Rick had marked. This time we were looking for species of tree-dwelling tarantulas, sometimes referred to as *Bird Eaters.* Though they occasionally caught birds, their normal prey are large insects. I told Rick about Henry Walter Bates's account in his classic book *The Naturalist on the River Amazon*...

> In the course of our walk, I chanced to verify a fact relating to the habits of a very large hairy spider of the genus Mygale, in a manner worth recording. The species was *M. Avicularia,* or one very closely allied to it; the individual was nearly two inches in length of body, but the legs expanded seven inches, and the entire body and legs were covered with coarse grey and reddish hairs. I was attracted by a movement of the monster on a tree trunk; it was close beneath a deep crevice in the tree, across which was stretched a dense white web. The lower part of the web was broken, and two small birds, finches, were entangled in the pieces; they were about the size of the English siskin, and I judged the two to be male and female. One of them was quite dead, the other lay under the body of the spider not quite dead, and was smeared with the filthy liquor or saliva exuded by the monster.

At midnight, on our way back to camp, we followed the course of a small nameless creek. I noticed lots of tiny eyes, like pin heads, reflecting in my torch beam. 'Rick, what are these?' I asked. As we moved closer to the muddy creek wall, tiny holes became visible. In their entrances were minute creatures. They were trapdoor spiders, no bigger than the head of a drawing pin.

These fascinating spiders excavate holes fractionally bigger than themselves and construct a flap of mud – a trapdoor – which, as they retreat into their tunnel, they draw up, sealing their home from the outside world. The doors mesmerised me. They were so *perfectly* formed to fit the hole that, when closed, it was impossible to see where they had been. I waited, my eyes inches from the surface of the mud bank, determined to see one open before me. After one shut tight I marked the spot just underneath it with a twig so I would know where to look. A couple of minutes later the tiny door eased free, and there it was. However, the find of the night was still to come.

Rick cannot walk normally. He maniacally scans leaves, fallen timber, roots, tree trunks and holes for insects and spiders. Utopia for Rick is a rotting fetid log – or a rainforest. He rooted the ground like a peccary most of the time – and it paid off. He found a spider as yet to be described by science. Although it excited us all that night, to put it into context it is estimated that there are thousands of insects and spiders as yet to be discovered in the Amazon rainforest. Nevertheless, it made our night. As I climbed into bed at two o'clock in the morning I wondered if it might be called *Avicularia rickus*. Then I heard the distant call of the 'poor me all alone' bird again and was soon dead to the world.

Our expeditions were not all centred around the rainforest! One Saturday morning, after breakfast, Jorge, always one for a good time, appeared and said, 'I thake you bothing today – thime for fun!' We travelled south on the bush road out of Ayacucho for 40 minutes, and pulled into a cleared forest edge on the bank of the Orinoco. There we found a lagoon that joined the main river a short distance away. Minutes later, another vehicle arrived, pulling the largest inflatable boat I have ever seen. Two giant bright red 25-foot tubes – just like the base of a bouncy-castle – were separated by five rows of seats, all fitted with safety harnesses. More alarmingly, there were also small harnesses on the floor for feet to be secured. While we puzzled over this contraption, Jorge informed us that we were going to experience white-water rafting through Raudales de Atures rapids. A wave of fear permeated every cell in my brain. Give me tarantulas any time.

The four people in the vehicle towing the boat joined Gordon, Rick, Jorge and myself for this jolly, white-knuckle ride. There was the boat handler, a pizza-bar owner from town and two friends of his. The fourth man, a dark-haired Venezuelan with a heavy black beard and thick round glasses, was a tour operator friend of Jorge's from Caracas. He strode over to me, hand outstretched with his business card. It announced 'Juan Carlos Lopez the Fifth!' I wondered if the other four had been lost white-water rafting.

Juan was very amiable, about 30 years old, and spoke excellent English. We all helped to push the raft into the water. It was silvery calm and my fears were allayed a little. We clambered in. Gordon and I had drawn the short straws and found ourselves in the front row. Sitting behind me was Juan the Fifth.

We motored out into the main river and then the skipper steered the raft into the bank while he showed us how to put our lifejackets on. He followed this with advice on the best way to hold on. We were strapped in so tightly, what was there to worry about, I wondered?

'In case we fleep over,' he said, 'you must release your harnesses and hold onto the safety ropes running along the outside of the tubes.'

Then I heard a distant dark noise. 'Gordon, do you hear that?'

'Sounds like a waterfall,' he replied.

We motored slowly with the current. I looked over my shoulder at the others. They all looked calm enough, then I saw the pizza man cross himself! No one spoke, and five minutes later no one could speak. That distant rumble had erupted into a roar. Suddenly we were in raging waters peppered with boulders, some the size of houses. The rapids spanned the full width of the river, about a quarter of a mile. The raft slewed sideways and nose-dived – I shut my eyes and yelled, 'What am I doing here?'

Gordon shouted back, 'Having a great time!'

Plumes of water spouted with no warning all around us. We lurched violently sideways, I thought we were turning over and shut my eyes again. All I remember was hearing a scream from Juan the Fifth, 'Holy *konia*!' and for the following 15 seconds we were under water. Fifteen seconds is a long time.

A glimpse of sunlight, a gasp for air, another scream from somewhere behind me and then holding our breath again, surfacing again. I couldn't let go even with one hand to wipe water from my eyes. Sensations that terrified and thoughts that made them worse ricocheted as violently as the raft around my mind. I'll kill Jorge if this doesn't get me first, I thought.

Suddenly we were becalmed, and I waited for the next torturous buffeting. There wasn't one. What seemed like an hour in hell had in fact only been 15 minutes. Looking back at the rapids I was beginning to laugh – we all were. I looked around at the others, drenched smiling faces, except Juan the Fifth who had turned white and was wide-eyed. I think he was still in shock.

'Why did you scream Holy *konia*?' I asked him. He told me that a wall of water simply rose up from the swirl right in front of us. It looked about 30 feet high and terrified him. Thank God my eyes had been closed. I asked him what *konia* meant? 'Oh, eetz Spanish for very bad word.'

'What word?' I pushed him to answer.

'Oh, eetz female anatomical,' he replied.

As my heart beat slowly returned to normal, Jorge asked if we wanted to continue to the port and do some water skiing. This sounded much more to our liking. Jorge had a speedboat moored at the port, and the pizza man dropped Gordon, Jorge and me there, whilst the others returned to camp. Julio had come to pick them up and laughed knowingly when he met us, still dripping wet.

Then Jorge sped us out to the middle of the river, about a mile wide. I jumped overboard, holding on to the side of the boat while Gordon passed me the mono-ski. We were drifting along at about eight knots. I gave the signal to pull, and in seconds I was skiing the glassy surface of the Orinoco. It was exhilarating – until I fell off. Jorge and Gordon didn't see me at first, then Jorge veered the boat towards me and the engine cut out. They were 200 yards away, both of us floating with the current at an alarming speed.

At this point I started to think about the Orinoco alligator. Jorge had the engine cover off and was fiddling about – they seemed a long way away. After a nerve-wracking ten minutes he got it going again and picked me up. I had had enough and soon we all gave up.

As we sipped ice-cold Polar beer Jorge said, 'Nick, when you get back to Caracas, better don't tell Charlie about your rapid experience.' Jorge then went on to explain how Charlie had brought an expedition from Caracas to the area and that their boat hit rocks as it went through the rapids. The boat had broken up and, tragically, 18 young boys had died. Charlie stayed in Ayacucho another month until all their bodies had been recovered, most of them trapped under the rocks downstream. Jorge finished the awful tale by saying, 'It was in 1972, I think. Better don't talk to Charlie about it.'

'You bastard, Jorge. If you had told me that before I wouldn't have dreamt of going for the ride,' I replied.

The following day my legs and arms ached. Gordon's face, arms and legs, along with mine, were vermilion red from sunburn. Caked in calamine lotion, looking like circus clowns, we returned to work in the film set.

Chapter Seven

A Slow Boat to the Lost World of the Yanomamo

After six months in Amazonia, we had come into contact with just three of the 20 or so groups of indigenous Indians living in Venezuela. The three tribes were the Yanomamo, the best-known internationally, the Penare, a shy race that seems to manage to combine trading in the commercial centre of Puerto Ayacucho with keeping to their traditional dress and customs; and the Piaroa, with whom we had been living and filming.

One of the strangest sights I saw in Puerto Ayacucho was that of a Penare Indian with a red-painted body riding a moped. He wore a lap cloth decorated on one side with a big pom-pom.

During our third month in Venezuela we visited a Penare village close to a Piaroa settlement. Almost everyone disappeared into their huts as we approached. They seemed a proud people. According to Jorge, they felt superior to other groups and they certainly take great pride in their appearance. 'Posers, compared to the others,' Gordon pronounced, watching a Penare man preening himself in the reflection of a local Ayacuchan shop window.

Shortly after we entered the village, the Penare chief appeared and asked us if we wanted to buy bead necklaces. They were beautifully crafted from forest seeds and tiny blue and white beads. I bought a bracelet for Emma made from red ormosia tree seeds and threaded together with a liana, a climbing vine, from which they extract a stomach medicine. We were told that the Penare had been the first Indians in the region to trade with outsiders and they certainly did appear to have come to terms with the world about them.

As far as we could tell, only small numbers of Piaroa Indians still lived in the forest, but many had been affected by so-called civilisation to one degree or another. Several villages we visited were in the early stages of contact with missionaries. In the worst examples, whole communities had been enticed to move out of the forest into settlements constructed by the government close to Puerto Ayacucho. Rows of small concrete block houses with tin roofs stood witness to the misery such policies created.

In one township, Gavilan, 40 or more of these bunkers had been built but behind many of the buildings the Indians had built their own traditional thatched dwellings, on a miniature scale. Privacy had now been forced upon them in the same way that they had been made to feel embarrassed about nakedness.

Here they were left to slash and burn, then farm the secondary bush around them. Some recreational facilities were provided which mostly stood unused and overgrown such as a football pitch, and at one centre, a modern children's playground. Julio told us that many Piaroans who came to Puerto Ayacucho ended up drunk, virtually penniless, lost in our world. From a race of people once completely in tune with their forest surroundings, and with their culture intact, they are fast becoming victims of our society, lured by false promises, a little money, and few prospects. A sad chapter in human history, a history that is still being made today – and not just with the Piaroa.

The Yanomamo have received widespread publicity in their struggle to survive, especially in the Brazilian rainforests. But what of their plight in Venezuela? The final part of our filming expedition to Venezuela was to take us deep into the southern part of Amazonas and into Yanomamo territory. Through Jorge, I had organised a 14-day journey south along the Orinoco and into the Casiquiare river. We would travel by dugout canoe with a fast outboard engine. This canoe was a kind of 'super' canoe: about 40 feet long with a small thatched canopy, which could travel at about 20 knots. Such canoes are quite common on the larger rivers in Venezuela and are called 'bongos'. We were going to pick up one of the smaller dugout canoes on the way so that we could investigate the tree-tangled creeks in the upper Orinoco area.

The main purpose of our river expedition was to find the Yanomamo and discover what relationship, if any, they had with *Theraphosa blondi.* We were to meet the bongo at Samariapo, one hour's drive south of Puerto Ayacucho. Here the bush road ended and our intrepid river journey was to begin. It was refreshing to see green forest all about us again, the environs of Ayacucho having been long lost to slash and burn.

We reached Samariapo by midday, despite the break-neck speed at which our driver Mario careered along, the trailer swinging wildly from side to side, several precious bottles of mineral water bursting as they fell out and hit the road. We picked up two English tourists along the way. They were both called William and wanted a lift as far as the National Guard post at Tama Tama on the Orinoco, about a five-day journey south.

Our captain was Florencia, a Piaroa Indian, who had great experience in the upper Orinoco. At this time of the year with the river so low, it was important to have someone with you who knew the water – or rather the rocks – well. He was about five feet tall, lean, but with powerful wide shoulders and walked as though he had been sitting on a horse for most of his 60 years. He was bow-legged, with trousers which were always held up by a bungy cord. He had a wart on his left eyelid that only appeared when he blinked.

We loaded our bongo to the gunnels with luxury supplies purchased in Ayacucho: soft drinks, dried bread, toilet paper and toothpaste. Our intention was to fish for meals on the way, camping at night on the sandbanks. By six in the evening on 22 February, 1992, we were setting camp on a sandbank a few hours south of Samariapo. We had stopped on the way at a National Guard checkpoint – a place called Islas la Raton, the island of rats – very aptly named! Five dubious characters in fatigues toting guns and dark sunglasses inspected our passports and permits. They confiscated the two cartons of tobacco we had brought with us to give to the Yanomamo.

The last hour of daylight was the worst time for the flies and they descended with a vengeance. It was a moody landscape with storms encircling us, the lightning playing along the horizon, and thunder rolling back and forth as heavy rain moved closer. Sandbanks are not the best places to be in torrential rain. But a golden glow from the fire warmed the air and while Gordon and Julio disappeared into the forest to cut poles to sling our hammocks on, I wrote my journal for the day, recording sightings of river dolphin, osprey, skimmers and many different parrots.

At first it seems a simple process to sling a hammock from a couple of poles embedded into the sand. The reality is very different. The poles, unless supported, fall inwards when you climb into the hammock. We learnt the hard way, of course, because Julio used to have so much fun watching us that he didn't tell us that the main poles needed guy roping, like tents, to prevent them collapsing.

It rained for a short while and we sheltered in the bongo, all hands messing about in the pitch dark of the night trying to sort out the hammocks. There was only room for three people to sleep inside, but luckily the two Williams wanted to sleep alfresco. After a lot of hole-digging to secure the poles, one of them leapt into his hammock, promptly pulling one of my posts over, depositing me on the sand. Enough was enough. Not wanting a confrontation, I derigged my hammock and moved into the canoe.

I awoke about five. It was remarkably cold. Despite the presence of piranha fish, the Orinoco felt lovely and warm as I plunged into it, but as I washed my hair the black flies returned and we were all bitten mercilessly. Julio had a fire going on the bank close to the bongo, and was boiling a saucepan of Venezuelan coffee. I imagined how lovely a bowl of muesli and fresh brown toast and butter would be... I pushed the thoughts away.

As the sun rose over the tree-line, the river looked enchanting in the warm early light of the day. Rick made a quick sortie into the forest to get his first spider and bug fix for the day while the rest of us struck camp and loaded the bongo. We were heading for San Fernando De Atabapo, and expected to have to haul the bongo for part of the way, this part of the river being very shallow.

When we arrived at San Fernando De Atabapo, we had to wait two hours for our travel permits to be stamped by another bunch of my favourite Venezuelans – the National Guard. Anyone wanting to travel in the interior of Amazonas – even Venezuelan nationals – require permission from the government. Next stop was to be Santa Barbara, and it would take two days to reach there.

The following night I saw clear starlit skies which seemed so close you really felt they were touchable. A shooting star briefly lit the heavens. A half moon was rising and reflected perfectly in the black, mirror-calm Orinoco waters. We cooked peacock-bass over an open fire on this perfect sandbank and then settled into our hammocks with a night-time view unequalled in all my travels. I gazed up, my whole vision the universe, galaxies and star clusters, all light years away, yet I felt I was in it and it was in me, body and soul. I did not want to go to sleep and tried so hard to keep my eyes open but eventually tiredness overpowered me and I sank into my own darkness.

At five o'clock I was awakened by a caiman alligator thrashing a fish to death at the water's edge. I indulged in another half-hour of star gazing until a dark red glow started to spread from the eastern horizon. The following hour almost equalled in splendour the previous night, with a sunrise so beautiful that I could not help grabbing my camera and shooting a whole roll of 35 mm film. Our hammocks were silhouetted against flecks of golden cloud while the thatched bongo was beached on the sandbar and mirrored in such a way that it was impossible to see where air and water met. Then a pair of river dolphins surfaced and dived about 100 feet out from the sandbank. Incredible images. Gordon heard my camera clicking and joined me, sharing silent enjoyment of the scene.

Before a mug of Julio's coffee could do its damage, we bathed and shaved. I noticed some tiny fish coming to the surface to peck at the floating shaving foam and Julio tried to put the wind up me by saying they were the dreaded candiru. The beast, according to folklore, enters the human urethra, or anus, and then erects spines so that it cannot be extracted without, in the case of man, amputation of the penis. 'Haay guys, look out for the dick eaters,' Julio crooned.

At 7.30 – mid-morning for us – we pulled away from the sandbank. Florencia always drove the bongo. He would sit at the stern on an old wooden fruit box, day after day. He was usually cheerful and smiling, although this morning he seemed out of sorts. Julio told me later, 'Haay, he's OK, he just found some worms in his shi-et this morning.'

Four black-headed caracaras arrived and circled overhead calling harshly. No doubt these scavenging birds would investigate the fish remains from our previous night's supper as soon as we had gone.

Our fourth day promised to be unbearably hot. Fortunately for much of the time our passage up the Orinoco was cooled by the speed of the bongo, with the small thatched canopy providing welcome shade. A black skimmer came close to our port side flying only inches above the water – hence its name. It is a fascinating bird with an apparent laziness of flight, flapping without hurry. It has a large blade-like red bill with the lower part of its beak, the mandible, considerably longer than the upper, the maxilla. When it comes into contact with food it quickly snaps the mandible shut. I have seen large flocks of these birds gathered on sandbanks, all sitting pointing their bright bills in the same direction into the wind.

By early afternoon we had reached Santa Barbara where we found ourselves once again confronted by the National Guard. This time, the five soldiers we encountered were with a group of four officials who turned out to be from the ministry of the environment. They looked out of place, dressed in pressed slacks and lace-up shoes, one even wearing a tie. Then one of the officials asked for our passports and Julio's notification of permission to be in the area. Julio turned to me and muttered, 'We gottus a problem here.' The official started to lecture Julio about how much damage tourists do, leaving litter everywhere and bringing disease to the Indians. Julio was looking extremely sheepish.

I was less receptive than I might have been, having seen at first hand members of the National Guard selling turtle's eggs, a protected species, to passing boats and enjoying target practice in the forest at night using a flashlight to pick out the animal's eyes and a rifle to shoot them. They also frequently used their position to extort supplies from passing traders. To hear this lecture on how bad we were made my blood boil. But after half an hour of diplomatic pressure from Julio, they grew bored with us and let us pass.

Later in the afternoon we pulled into the mouth of a beautiful narrow river curtained with untouched forest. At first glance, the trees all look the same but soon you realise you can walk hundreds of yards or even miles before seeing two trees of the same kind. This is what diversity is all about.

Florencia told Julio he wanted to show us how to catch piranha fish and we tried to muster our enthusiasm. I didn't have the heart to tell him that we had survived for months by catching the things ourselves. They are such voracious creatures that fishing for them is easy. You just have to be extremely careful how you remove them from the hook!

Florencia prepared his tackle, the final ten inches of line being a metal trace. If you don't use metal, the fish will chew through the nylon. Within half a minute he started pulling piranha in. Using a paste made from manioc for bait, sometimes the hook had only touched the surface of the river when a fish snatched it. The most dangerous to deal with are the large black piranha. Weighing several pounds, and with rows of teeth like saw blades, getting one of them off the hook is a risky business – as Florencia soon found out. The fourth fish he pulled into the boat was a black piranha, and as usual he smacked it hard with the blade of his machete. As he grasped the wet slippery body behind the head, and tried to wrestle the hook from its mouth, it lunged forwards grabbing the knuckle of his thumb. He howled with pain as blood flowed alarmingly. Seconds later he was howling again, but this time with laughter. His lesson in how to catch piranha had gone sadly wrong.

I spotted a single male giant otter. He was bobbing up and down in the water to get a better view of us and we shut down the engine to enjoy a few minutes watching him. We set camp close by on a small sandbank, surrounded on three sides by tall forest. It was a wonderful place to stop and rest, that is until suddenly we were surrounded by swarms and swarms of tiny black flies, each leaving the telltale speck of blood on the skin. Darkness brought relief, but also brought high winds and rain.

The rain did fall, and heavily, but only lasted an hour and a half. We all huddled together in the leaking shelter of the bongo's roof, sleep an impossible dream as water dripped all over us. Gordon drew the short straw and disappeared into the forest to cut poles to hang our hammocks while Julio, against all odds, made the most delicious hot fish soup using the afternoon's catch of piranha. It was almost midnight when we all fell into our soggy hammocks, enveloped by high-pitched whines from the mosquitoes.

The dawn brought a marvellous chorus from the as yet invisible birds in the trees around us. We didn't have to wait long before the birds took to the air, giving us magnificent views of scarlet, blue and yellow macaws, channel-billed toucans, a white-tailed trogon, hundreds of cackling parrots, a screaming piha, woodpeckers, hawks and more. The sightings lifted our damp spirits until the flies returned. The two Williams went for a jog along the sandbar but the flies soon put paid to their exercise. They looked so incongruous running along the Orinoco sandbank in their designer outfits and Panama hats.

We struck camp and, within minutes of rejoining the main channel, saw a family of giant otters and a river dolphin nearby. The large male was eating a big, scaled fish on a half-submerged log and did not seem too perturbed by our presence. We watched for 20 minutes until they crossed the river together, peering at us from behind a fallen tree, and finally disappearing from our view into a creek.

Back to the mundane, a sore on my right hand was giving me some discomfort – it's odd how it's always the trivial details that seem to cause problems. Digging the holes for the hammock poles the previous night with my bare hands had rubbed the skin off the edge of my fingers and the wounds were now turning septic. This is a typical reaction to any bites, cuts or grazes in this climate where the dampness and humidity conspire to slow down the normal healing process.

By mid-morning we were on the go again and passing Yapacana, known as the Golden Mountain. A magnificent *tepuy,* it towers above the surrounding forest and was topped with a huge cumulus cloud, which gave it the appearance of a giant smoking chimney. The area around it is rich in gold deposits and although a protected region attracts many illegal mining operations. We stopped for an hour to fish. No sooner had we jumped onto the rocks than another giant otter showed itself, periscoping and bobbing to get a better view of us before making his way slowly up creek and out of sight. As we motored off, a flock of about 50 blue and yellow macaws flew noisily by and into trees very close to us. With my binoculars just a few of them filled the lenses' vision.

We stopped a little way further up-river to fish, still in the shadow of the Golden Mountain. Julio managed to get a good catch of piranha and two large peacock-bass, which to my mind taste like halibut, to boost our meagre food supplies for the next two days, while one of the Williams sucked tangerine segments in a ritualistic sort of way, one by one and very slowly. Two hours later we arrived at a small Indian settlement called San Antonio. There were about eight thatched huts, where 35 or so people lived. The houses were all built close together and looked dilapidated, several with large holes in the roofs. A small peccary (forest pig) and several chickens were scrabbling about in the dirt in front of us as we walked up to the door of the nearest hut. The inhabitants were small in stature, probably not more than five feet tall, but with attractive faces and jet black hair. They looked poorer and more dishevelled than the other communities we had visited, with tattered clothes. Clearly no missionaries had visited here for a long time.

They offered us a partly-built thatched hut for the night, and, with darkening skies, we decided to forgo the sandbank and for a change sling our hammocks under shelter in a hut.

We retired to our hammocks early after a delicious fish supper, which we ate on a rickety table. It was cooked by a very pretty Indian girl. The candles were doused, howler monkeys serenaded us, and a dog crept in and stole our last bag of cheese crisps.

The sound of heavy rain woke me. Thank heavens we opted for shelter tonight, I thought, while I watched a stream of rainwater fall directly over Julio's hammock. We spent a lazy morning lounging in our hammocks, enjoying the shelter, as the rain continued and thunder rumbled all about us. We heated some coffee and ate hard-boiled eggs between stale bread, with pepper and mayonnaise added to give it some taste. What I would have given for fresh orange juice and grilled Manx kippers!

The rain petered out by mid-morning and after thanking our hosts we set off on the last leg of the journey to Tama Tama. We passed two small mountains and Florencia told Julio that the first was called Mountain of the Virgin. It transpired that this was so because he had deflowered a young native girl there many years ago. Predictably, everyone howled with laughter as the usual jokes followed about whether Florencia had anything to do with the naming of the second one – Tapir Mountain!

We huddled together under the bongo's canopy, dreary from the wet and cold. Our spirits perked up when we spotted a flock of hoatzins. The adults' faces are bright blue and bare of feathers, and their eyes are red. Capping off this unlikely looking bird is an untidy fan-shaped crest. Their Portuguese name is *cigana* which means gypsy. I counted more than 20 of them as we moved in for a closer look. Hoatzins are usually found among bushes and low trees at the edge of remote forest streams and rivers and they feed purely on vegetation. They are not very skilled aerobatically as they use their wings mostly for gliding; we watched several crash-land into branches. They are noisy and according to many Indians have an unpleasant smell – in some areas they are known as the 'stink bird'. When their naked chicks hatch, they n ancient biological feature – two claws at the bend of the wings, re used by the young to climb about the bushes. They are excellent ers at this tender age and should they fall into the water they can and pull themselves out with the help of their clawed wings. Both the s and the swimming ability are lost two or three weeks after hatching.

n the last minutes of daylight we pulled into Quaraeare, a small tlement inhabited by Warekena Indians. Some of the Indians were at the water's edge, cleaning a good catch of catfish. A local family prepared dinner for us and I had my first taste of a meat that I would normally have refused – tapir. I couldn't liken it to anything such as pork, beef or even game. It was incredibly salty and tough – and was no gastronomic delight, that's for sure. I would much rather have seen it left alive. Tapirs are becoming quite rare.

We ate the meal in the family's hut, which was divided into two parts separated by a partition made from woven palm leaves. The eating area was lit by a small oil lamp fashioned from an old food tin. A roughly-hewn wooden table and two benches were in the centre of the earth floor. There were no windows, and the atmosphere was heavy with humidity. An old man with dreadful sores on his legs sat tying fish hooks onto a line while a pretty young girl, about 15 years old, carried the food remains away. One of the Williams proffered, 'If she scrubbed herself up I could give her a good seeing to.'

'Colonialism still exists then?' Rick asked him, as he went out of the hut leaving William to his fantasies.

After dinner, I foolishly agreed to a game of poker dice with the two Williams. My other travelling companions largely ignored them and although the Williams never did anything to endear themselves to us, I sometimes felt I ought to make an effort to include them. You don't meet many fellow countrymen in our business, and if we were all the same it would be a terribly boring world.

We set the penalty of losing a game as taking a shot of neat rum. After losing the first four hands my resolve went out of the window, if there had been one, and four hours later we three were squat on the earthen floor, talking utter nonsense. The following morning's headache required two Neurofen and the only saving grace was that, judging by the moans and groans from the two Williams' hammocks, they were in a far worse condition. As I made my way down to the river, Florencia stood with two of the Indians, howling with laughter, mimicking the night's drinking. '*Enratonada*,' he called out, which apparently translates as 'mice running around in your head'. We pulled into the most glorious spot on a sandbank to wash the 'mice' away. The cool, fast-flowing waters refreshed us all and after making a fire and boiling coffee we made our way up river and, finally, late that afternoon, into Yanomamo territory.

A meeting with the anthropologist Napoleon Chagnon in Caracas came to mind, especially a story he had told me about his first visit to a Yanomamo community – by helicopter. The Indians had gathered around the machine as it was landing, not understanding the danger from its rotor blades slicing through the air above their heads. The pilot was waving furiously at them to move away. As they climbed out of the helicopter the Indians chattered excitedly while circling it. Chagnon, who can speak their language, told me they were trying to find out what sex the bird was!

As agreed, we dropped the two Williams off at the National Guard post of Tama Tama, where they would be picked up by light aircraft and flown back to Ayacucho. Tama Tama was little more than two cement huts and four guards. Our mood lifted with the Williams' departure although I pulled Rick's leg suggesting the aircraft might not show. He didn't find it amusing. Julio made some remarks about all Englishmen being the same and, of course, Rick and Gordon backed him up, laughing their heads off. But everyone stopped laughing when the bongo ran out of petrol and Florencia poured the last four gallons of fuel into the tank, telling us we might run out before we reached Cejal, our destination. He planned to return to the airstrip to get gasoline in the morning. I didn't like the idea of us being stranded for a day or more without a boat.

We passed a Yanomamo settlement where two women were sitting on rocks. I looked through my binoculars and could see their faces quite clearly. They had several thin sticks piercing their noses and lower lips. Only minutes later a naked Yanomamo man paddled past us in his dugout canoe, loaded with plantains from his forest farm.

We headed away from Tama Tama into the Casiquiare river to do some fishing, stopping at a bend where large, dark rocks swept down from the forest edge into the water. The rocks were covered in ancient carvings known as petraglyphs, the first that I had seen. I crouched down to look at the shapes of people, lizards, monkeys and other designs and suddenly heard high-pitched squabbling noises. They appeared to be coming from a thin crack in the giant slab of rock I was standing on. Crawling on all fours with my torch revealed a colony of 20 or more tiny bats that were spending the hours of daylight roosting in the most inaccessible of spots. Their faces twitched and tilted as they tried to work out who this intruder was.

We arrived at the Yanomamo community of Cejal an hour before dusk. They immediately impressed me as friendly people. We were met at the riverside by two small men. The younger, about 20 years old, told Julio he was a school teacher from a distant village where missionaries had set up station. He spoke good Spanish. The older man was silent. His bottom lip was grossly protruded by a thick roll of black stuff that looked like tar but was in fact tobacco. Dark dribble stains lined his chin. They offered us a place to sleep for the night.

When we met the Yanomamo chief, a short, powerfully-built man, we soon discovered that a fight was simmering between his people and another Yanomamo group at a settlement called Ocamo, about a day's river journey away. The fight was over a woman, a common situation in their culture, I was told. The high ratio of men to women leads to the Indians mounting raiding parties to kidnap girls from other communities. These raids then have to be avenged. They were expecting an attack on their village either that night or tomorrow. 'Here we go again,' I said to Gordon.

A group of girls and women were bathing and preparing plantains in front of me, their faces pierced with wooden ornaments above and below the lips. These were white sticks about six inches long and as thick as drinking straws. Some of the women also had shorter ones through the soft centre skin at the base of their noses. These are called praiai, and the girls are prepared by their mothers to take these from the age of one, along with both ear-lobes. Boys have only their ears done at the same age.

A young boy was squatting by the river's edge gutting catfish, using his teeth to descale and rip the flesh open. The night was dark, with no moon. While Rick went into the forest to search for spiders, I sat with the chief and, with Julio translating, talked about the tarantula and the Yanomamo's use of it. They call it *haho* and only eat it as a cure for someone suffering from bad dreams. Unlike the Piaroa Indians they appeared to have no spiritual beliefs concerning it.

After our talk the chief excused himself and went into one of the houses. We heard some strange half-human noises coming from it. An hour or more later we could hear someone moaning as though in great pain. Another dwelling was the source of strange chanting. We never found out what it all meant but it must have been the prompt for the strange dreams I had that night. The atmosphere here was beginning to disturb us all.

All morning there were strange comings and goings. A growing tension was heightened by further incantations coming from inside one of the larger long-houses.

Florencia had left in our bongo to motor further up river to a settlement to find some gasoline and during the afternoon Rick and I were walking between two of the thatched houses when we were unexpectedly ushered –

in fact virtually pulled – through the entrance to a long-house by a young Yanomamo man. The muffled chanting and cries that we had heard walking up from the river were now filling the dust-laden air. The interior was humid and wreaked of an odour familiar to me – *yoppo.*

It took a minute for my eyes to adjust to the darkness. The chief sat in the centre along one side of the hut's wall. Facing him, forming a semi-circle some ten feet away, were 18 Yanomamo warriors. Behind them were other men who were obviously preparing for a fight. Long arrows were being flexed menacingly and thrust into the air. In one corner three young men were putting the finishing touches to their formidable weapons. We were extremely frightened. I couldn't understand a single word and was wishing hard that Julio and Gordon would come looking for us.

On the bare earth between the chief and the warriors the *shaman,* a strong-looking medicine man, was in full flight. He chanted loudly while moving dramatically around his arena. He mimicked with slow ballet-like movements an act of Yanomamo war. These normally friendly people were preparing for a battle. There was no doubt that they meant business.

Later, one of the young warriors told us that they had moved to this site 11 years ago from a community far up the Casiquiare river. There had been a fight among their own villagers and many people were killed. Those that survived and fled took three months to walk through the forest to their present village, Cejal. Only a few days before we had arrived the Cejal villagers had mounted a revenge raid on another group at a village called Ocama. They had killed three men. Apparently, a year before, some men from Ocama had stolen a woman from Cejal.

The chief ordered Rick and myself to squat down, communicating with hand signs. All the warriors were painted with a dark red dye on their faces and bodies. Wild, staring eyes occasionally looked at us. The *shaman* moved quickly towards me and held his palms out flat in a kind of stop sign. His eyes unnerved me. Each warrior was summoned in turn by the chief to receive a blast of *yoppo.* A three-foot long pipe was loaded musket-style with the hallucinogenic powder which the warrior held up one nostril while the chief blew long and hard down the other end. The powder came out of the warrior's other nostril, his mouth and in some cases, it seemed, his ears. Reaction was instantaneous. Within a minute we were surrounded by warriors throwing back their heads, clutching their skulls with one or both hands and vomiting on the earth floor.

All the men were getting more and more intoxicated, having done this for the past few hours, and tensions were rising. Rick's left leg pushed against mine, our eyes met briefly, but we dared not even speak to each other. The *shaman* danced, cried and shouted. My first fear was of the *yoppo,* and sure enough it was not long before the pipe was being offered to me. How not to offend? I knew from my experience with the Piaroa Indians that at the very least I would be sick, but the Yanomamo method of delivery was very different and I did not want my ear drums blowing out, as theirs presumably had been. I used sign language to decline and hoped they would understand. Rick made the same sign of refusal.

The *shaman,* still chanting, went to one of the older men and began a strange sequence of movements over his squatting form. He was gushing mucus from his mouth and nose onto the floor at his feet. The *shaman* stroked the warrior's torso with his hands to give him protection from evil spirits. This man had a devil within his body that needed to be removed. He put his mouth to the man's right ear and began sucking hard. After about 20 seconds he pulled away and we heard a pop. The medicine man began to chant again and then convulse. His stomach muscles pushed upwards and then he vomited. Putting his hand into his mouth he pulled out what looked like a sliver of bone or wood and this he took over to the elders for confirmation that this was indeed the evil spirit he had sucked out of the man's body. He then turned his attention to the man's left shoulder, and after that, his right knee, repeating the whole process from chanting to vomiting violently. While all this was happening, the others continued to take more *yoppo* and the young warrior to my left threw up, splattering my left leg and foot. Neither Rick nor I dared move a muscle for fear of being noticed. The deep privilege of seeing this was no mitigating factor to the fear of attracting attention to ourselves.

The weapons normally used for hunting and warring are long spear-like curare-tipped arrows. These and the bows are more than six feet long. I saw some of the warriors making much smaller versions of the same thing – we were told later that these were for the young male children of the community to use when they were attacked. The chief's ten-year-old son, Cagi, said to me, with Julio and the young school teacher interpreting, that his father was very angry and wanting to kill. It was strange to be in the presence of such a young child, only a little older than my own daughter Emma, talking about killing people. He was even smiling as he said it.

Julio eventually arrived at the long-house. Although by this time we were as near to flies on the wall as we could be, both Rick and I were relieved to hear his voice calling for us outside. We rose slowly, and without looking at anyone moved out through the small entrance. It was only as we walked towards the river that we both realised how nervous we had been. We burst out laughing, almost disbelieving what we had just experienced.

'Tell me, guy, did that really happen?' Rick asked.

'Let's get down to the river edge,' I replied. As we reached the rocks, 12 of the *yoppo*-crazed men appeared from between the houses, shouting war cries. Their red paint had been replaced with black, which signifies real war. They ran into the high forest to take up guard positions around the small community.

How those men could co-ordinate their body movements after so much *yoppo* I would never understand. I sat on a slab of rock at the water side to write up my journal and prayed for our bongo to return sooner rather than later. Our plan was to board it immediately and then go down river to a safe sandbank for the night. We all knew that we could not stay at Cejal.

A fever-pitch of expectation was building up. I watched some of the older women, normally casually slow in movement, strutting about stiffly, presenting an air of aggression as they walked past each other. I even saw one girl, about nine years old, lash out at a passing dog – she was as worked up as everyone else. The warriors' howls and whoops could occasionally be heard from the forest and the women and children echoed the calls.

During the final hour of daylight the women and children came down to the river to wash. Each group was accompanied by one of the warriors who made a point of standing on the summit of the largest rock and posturing with his bow and arrows, then drawing the bow back and pointing it skywards. Any enemy watching was left in no doubt that the men were there and armed. These activities were punctuated by frequent choruses of screams from everyone keeping the frenzy on the boil.

It was nine o'clock when we finally heard the sound of the bongo's engine through the still night air – never was such a din more welcome. By the time Florencia pulled the boat in at the rocks we were already there with all our equipment, ready to board straight away. We didn't even say goodbye to anyone.

It was slow progress in the dark. Julio decided we should all spend the night at the National Guard post back at Tama Tama. Not a soul was to be seen when we arrived and we all slept on hard ground.

By 6.30 the following morning we were boating north, having decided the only sensible thing was to head back to Puerto Ayacucho. We estimated it would only take three or four days, going with the current. Our brief time with the Yanomamo had been anything but dull. Safety concerns apart, we had learned about our tarantula and we still had much work to do in Puerto Ayacucho.

At mid-day we pulled in to some rocks at the confluence of the Orinoco and a beautiful, black water creek called Trucuapure. It was time to wash the grime away and – for the first time in days – brush our teeth. As we clambered out of the canoe, we saw a family of three giant otters emerge from overhanging vegetation and make their way up the creek, fishing and resting. Then, as if an artist were putting the final brush stroke to his masterpiece, a scarlet macaw flew close overhead, calling loudly.

While trying to get a photograph of the two dolphins and the giant otters, with the creek stretching away in the background, as usual my excitement got the better of me. I leapt from one high rock onto a large boulder at the water's edge, but to my horror the rock was so steep I could not stop. As I hurtled forwards, realising that there was nothing but the deep creek to catch me. I thought only of my £2,000 worth of camera gear. I decided in a split second to jump as if off a springboard, extending my arm high above my head to save the camera. Gordon and Rick turned around as I called out, and saw me mid-air laughing at my own predicament. I did manage to keep my equipment dry, although I was submerged. My three companions erupted into paroxysms of laughter.

The remainder of our passage north was a relaxing contrast to what had gone before, but not without its memorable moments. At a place called Babilla, the waters of the Orinoco – about a quarter of a mile wide – gave us another wonderful wildlife sighting. I was dozing when Rick suddenly shouted he could see something crossing the river ahead. It turned out to be a superb specimen of the largest snake in the world, the anaconda. And it was a big one, about 18 feet long. We followed it across to the opposite bank where it came out of the water in full view and rested for a moment. What a beauty! It was placid and unconcerned as I jumped out of the bongo and crouched within two feet of its head. Florencia was extremely nervous, telling us to get back into the boat at once – it could jump and was poisonous. We tried to explain that the anaconda was not a venomous snake and could not jump, but Florencia insisted it was only 'not poisonous' in May. This was when someone he knew had been bitten and survived. The anaconda eventually swam away under our canoe, leaving us thrilled by the event.

That evening we set camp on another sandbank of the Orinoco. Within a few weeks they would be covered again by the rising waters. Hammocks were strung and Julio tried to catch some fish. We were very short of supplies – we were left with the sum total of a little sugar and coffee, two tins of sweetcorn, a tub of margarine, cream-crackers and four eggs that Julio had begged from the last community we passed. There was no danger of us starving but it did bring home how much we take food supplies in the West for granted. That night Gordon and I dined on egg mixed with sweetcorn and tomato sauce. It was strangely delicious.

We made Santa Barbara by sunset the following day and, by pure chance, learnt that a bush plane to Ayacucho would be leaving the next morning. Gordon, Rick and I decided to go for it and avoid meeting the 'rats on Islas la Raton' again, while Julio and Florencia continued by bongo. Our exit from the area was fitting, everything considered. The bush plane rattled in, complete with rubber bands, and we piled aboard. Nobody expressed any concern about our weight or the weight of our gear. The six-seater lumbered along and gradually lifted off – only then did the pilot mutter something about how heavy the plane was. Then, to add to our nerves, while climbing through the 1,000-foot barrier the pilot started swatting at a stowaway mosquito. He removed both hands from the controls to pursue it, then tried with his aviator's map, while we bounced about in thick cloud. Finally Rick swatted the beastie from the co-pilot's seat. With perfect timing, our bush plane came out of the cloud, giving us a sudden snatch of the Amazonas forest and the sacred mountain of Wahari Kuawai, where our quest to find the Piaroa Indians and the giant tarantulas of this lost world had started.

Take Two: Brazil

Chapter Eight

Across the Frontier

Jorge met us at Ayacucho airport on our return from our Yanomamo expedition with the news that an urgent fax had come from England. Survival wanted me to go into Brazilian Amazonia to reconnoitre an area where our next film might be made. In that instant I thought of the man who epitomised the word 'Hero' for me as a schoolboy. Henry Walter Bates wrote what is probably the greatest natural history book ever written, *The Naturalist on the Amazons: A Record of Adventures, Habits of Animals, Sketches of Brazilian Life, and Aspects of Nature under the Equator during Eleven Years of Travel.* Published in 1863, it is a classic I first read as a 14-year-old. As a 44-year-old, it is still never far from my reach. Could I really be going to the land that so captivated Bates 130 years ago?

Survival had been approached by an organisation that called itself the Amazon Association with suggestions for an area to film a Survival TV Special. For many weeks I had been discussing with Jorge the prospect of settling in this area for a few years to make a series of rainforest films. I wanted a base – a home even – that would give me the continuity I lacked. Amazonia was where I wanted to be, so I was on the lookout for just such an opportunity. Little did I know that I would eventually settle in Amazonia for at least five years, build a permanent home in the forest, and make four films for Survival.

But that is jumping too far ahead.

Back to the present. Gordon could arrange the final sequence linking shots to finish the giant tarantula programme. I would go to northern Brazil for three weeks while he would have a quiet time in Venezuela, or at least that is what we thought. Three weeks is a long time in South American politics.

On the 24 January, 1992, I flew to Manaus, Brazil. The city itself sprawls untidily at the junction of the Amazon River and the Negro, about 1,000 miles inland from where the giant river pours into the Atlantic Ocean. I had heard Manaus described as the 'city in the jungle'. Looking at my map, it was a fair description. The only way into it was by air or boat – unless one was prepared to risk the Boa Vista track which, on the map at

A darter or anhinga (Anhinga anhinga) *drying its wings after fishing.*

A jaguar (Panthera onca) *with its freshly-killed tracaja turtle* (Podocnemis tracaxa).

A blue and yellow macaw (Ara ararauna) *– just one of the many, magnificently-coloured birds in the rainfor-*

Xixuau ('rainforest') Creek – remote, wild and beautiful.

ABOVE *Getting the best shots took me into some uncomfortable situations.*

LEFT *The piranha fish that nearly bit Luiz's finger off!*

BELOW LEFT *A tapir surfaces. The meat of these rare animals was served for dinner by Warekena Indians when we stayed with them.*

BELOW *One of my favourite pictures, a giant otter* (Pteronura brasiliensis) *eating tucanare fish. The river in the background is the Orinoco.*

ABOVE *Our host, Carlitos, drinking from a water-filled vine.*

LEFT *A Yanomano girl. A great people – and great faces. She was laughing at us when this shot was taken.*

Carlitos' ramshackle home.

Francisco hunting fish with a spear – despite his tender age, he is an expert fisherman.

ABOVE *Gordon and me – as usual, up to our necks in work!*

LEFT *We spent many days filming the flooded forest – I was always wondering where the two black caiman were lurking.*

BELOW *Filming the magic waters, this contraption (a micro-focus jib) allowed us to take shots above and below water.*

ABOVE *The anaconda* (Eunectes murinus) *that swam between my legs was roughly this size – no laughing matter for me! Holding the snake are camp crew Antonieta, Stephan and Almir.*

RIGHT *The amazonian manatee* (Trichechus inunguis) *– a giant and rare vegetarian who has found refuge in the Xixuau.*

BELOW RIGHT *A mata mata terrapin* (Chelus fimbriatus) *swimming underwater.*

A flock of roseate spoonbills (Ajaia ajaja) *adds to the wild beauty of the Rio Branco.*

Tracaja terrapins queuing to get out of the river to bask in the sun.

This tambaqui fish is just one of the many varieties we found in the flooded forest.

least, joined Manaus with the borders of Venezuela and Guyana. I knew people who had tried that route and had had to be rescued from the middle of nowhere by the military. The 'road' is simply unpassable for half the year.

I was met at the airport at 3.00 am by Chris Clarke, an Englishman whose home is in Italy. He was in his early thirties, about five feet seven inches tall. He and three partners, Daniel Garibotti, Plinio, and Eric Falk, had set up the Amazon Association, he told me, to create an ecological reserve in the Rio Xeruini, a tributary of the Rio Branco. He took me by taxi to the port of Manaus to board a boat that would leave immediately for their reserve. He spoke in glowing terms about the wildlife we were going to see. I was already excited.

From the little I could see, Manaus was a shabby city. Even at this ungodly hour many people spilled across roadways and loud music buffeted our car as we passed shacks that Chris told me were nightclubs.

'Do you speak Portuguese?' he asked. 'No, but I'm a quick learner.'

'Good, because the local people don't speak a word of English,' he added. We passed a huge billboard that announced '*Sexo Explícito*' – it was a cinema, and I had learnt my first two Portuguese words.

The taxi went through two enormous wrought-iron gates and passed between dark metal mountains made of ship's containers. We bounced and rattled over a bridge linked to a huge floating pier. A metal plaque announced that this dock section was built in Glasgow at the beginning of the century. All around the edges the bustling pier was clustered with all sizes of colourful wooden riverboats. One three-decked vessel was crammed with hundreds of brightly coloured hammocks, like tinned sardines.

Chris pointed at one of the smaller boats and our driver pulled up in front of it. The name plaque said *Commandante Barata* – Captain Cockroach. It seems I could not escape the things!

Barata, rented exclusively for our 3-week recce, was about 50 feet long and 12 feet wide, with two deck levels. From the bow a small ladder was fixed to the front of the upper deck. A group of about ten people were huddled there, all looking at me. As I climbed aboard, I passed my rucksack and camera bag to a wizened stooping man who turned out to be the owner of the boat – and his nickname was indeed 'Cockroach'.

'Welcome American,' he said.

'I am English,' I bridled as he scuttled off ahead of me.

Chris presented me to the others as the boat pulled away from the dockside. A tin of cold beer was pressed into my hand by one of the crew members, Justino, a rotund, happy-looking man. His permanent grin showed a row of gums and three tobacco-stained pegs, one that wobbled alarmingly as he talked. He was a Caboclo, a people of mixed Indian and Portuguese origin who have settled the river edges throughout most of Amazonas. A tall, good-looking man stepped forward offering his hand. Daniel Garibotti was an Argentinian, an English speaker, and a partner in the Amazon Association. He was friendly and welcoming and I liked him immediately. In his shadow was a small, powerfully-built young man. 'This is Plinio,' Chris said as I put my hand out. He was the president of their association. He shook my hand firmly, but avoided looking into my eyes. He was a Caboclo, about 26 years old, and had known Chris and Daniel for eight years.

Then there was João. He was the *mateiro*, or forest expert. He shook my hand while playing an old mouth organ. There were two Italians, Lauro and

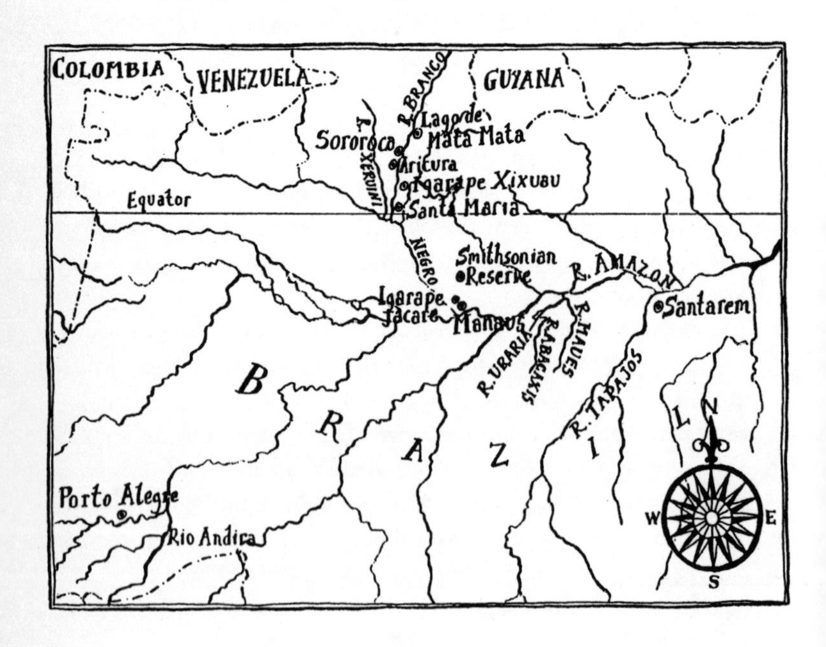

Antonio, who were going to be dropped off at some point to live for three months in the forest, surviving by hunting. And, finally, Marciel.

The upper deck was cooled by a breeze as we motored away from the lights of Manaus. Even in the dark I could see that the river Negro was vast. I slung my hammock under the upper deck canopy next to Daniel. By torch-light, he showed me a map and traced the route we would take to the Rio Xeruini, and then up the Rio Branco to the turtle nesting beaches at Sororoca, in total about 750 miles. Of my 21 days' recce, 18 would be on the boat and the last three back in Manaus meeting the institutions that could help – should I decide that a film could be made here.

Despite the constant pounding noise from the boat's diesel engine, and João's love affair with his mouth organ, I fell asleep quickly. I was woken just before nine by Justino, who had been told to look after me. He was waving another tin of cold beer above my head. I politely refused, using sign language by prodding a finger on my watch to say 'too early' but his smile and puzzled expression told me he didn't understand.

A Caboclo woman called Fatima was our cook. She was about 35 years old and was setting out the breakfast table. '*Bom dia senhor,*' she said. '*Bom dia,*' I replied, mastering the obvious. She flopped a dozen greasy, fried eggs onto a plate. One slipped off, hitting the diesel-smeared wooden deck. She spiked it with a fork and put it back on the plate, smiled suggestively at me, then placed them on the table next to a platter of cheese and ham! I tried to memorise the rogue egg's position for later. As I passed Fatima, she patted me on the bottom. I laughed and she joined in. She could not have been described as shy!

The bathroom was built for Caboclos, by Caboclos – a race generally little more than five feet tall. There was a shower head on the ceiling and as I washed I laughed to myself at what I must have looked like, leaning over backwards as though about to do the limbo.

The next day I woke up at 4.30 am. We had pulled into a sandbank the previous evening and now, with everyone else asleep, I climbed over the edge of the boat to investigate. As I looked back at the boat, the sky behind it was black but the horizon had a dark red hue to it. I walked right along the edge to where the sand met water, scanning the ground with my torch. I spotted a large stingray slowly sifting the sand for food. Its 'wings' undulated, and the venomous spine near the tip of its tail was clearly visible. I thought of Pegleg in Venezuela and what that spine had done to him. Several times I stopped to listen to a dolphin surfacing close by – but each time I looked all I could see were the spreading rings on the surface, I marvelled at the strange sounds it made as it exhaled hard and then inhaled, noisily sucking in the air, before diving again.

My torch picked up the tracks of an animal that had come to the water's edge during the night. Then I saw that the paw marks moved off in the opposite direction – towards our boat. It was a jaguar. Suddenly I realised how isolated I was. I was at least 300 yards from the boat, with no cover and no trees – the forest was almost a third of a mile away. I didn't believe that jaguars attacked humans unprovoked, unless driven by extreme hunger, but right now my reasoning was giving me little comfort. I saw a torch-light on the boat and started to head back towards it, all the time peering warily into the half light. In my mind, predatory shapes jumped from the shadows of the distant forest edge. To my relief I saw the big cat's tracks veered off right, away from the boat and towards the forest.

We set sail at 9.30 am after an egg, cheese and chilli sauce 'butty' breakfast. I settled on a bench at the back of the upper deck, my binoculars around my neck. Within the first two hours I had seen a magnificent assortment of birds: scarlet macaws, anhingers – diving birds like cormorants – and a falcon that swooped out of the sky, plucking a blue morpho butterfly the size of a saucer from just above the water's surface. The area was virtually uninhabited – all the signs were looking good.

After nine o'clock in the evening we arrived at the mouth of the Rio Jauaperi, and the Caboclo home of Valdemar and his family. Chris, Daniel and Plinio had known him for some years, and there was much mutual back-slapping. He was about 30 years old, and despite his youth had seven children tottering about the veranda of his wooden hut. His wife was breast feeding an eighth, a newborn baby. Fish was spitting away on an open fire but I was not to bask in such home comforts for long. Daniel informed me that Valdemar wanted to take me for a paddle up the nearby creek to show me the night life, and we were still searching creeks until after two in the morning.

Valdemar was an expert at paddling the dugout canoe silently. He must be a good hunter, I thought – and he would need to be with all those mouths to feed every day. As he shone my torch to check the water's edge many pairs of red eyes stared back at us. They were caiman, a type of alligator. Valdemar told me that the formidable *jacaré açú*, or black caiman, which grows to an awesome 21 feet, lives in these waters. We weren't lucky enough to see one this night, only these six-foot tiddlers. We also saw a tropical screech owl, perched on a low branch over the water. Like a spectator at a tennis match, caught by the camera in slow motion replay, its head and big yellow eyes followed us from right to left as we paddled by.

On the way back to the *Barata* we had our final sighting of the night. Judging by Valdemar's excited gestures it was something he didn't see too often, an ocelot (a type of cat), and jaguar tracks within 24 hours – not bad at all!

The following morning I was woken at 5.30 am by the sound of our boat's diesel engine. An hour later my hammock began to sway so violently I felt in danger of being thrown out. We were crossing a wide stretch of river that would bring us to the mouth of the Rio Branco, and the waves caused the boat to rise and fall sharply.

The river edge and sandbanks here were teeming with wildlife. As I hungrily demolished a less-than-romantic egg and cheese sandwich, smothered with Tabasco sauce, I watched ibis, herons, woodpeckers, ducks, parrots and oropendolas take to the air. The only thing missing was a large mug of black Earl Grey tea.

I noticed something odd as we sailed along this stretch. The water on our port side and underneath the boat was dark, almost black. Yet to our starboard it was the colour of creamy coffee. The contrast was so sharp it was baffling. Then we realised we were seeing the meeting of two great rivers here, the Branco's white water and the Negro's black. Indeed this is exactly what their names mean in Portuguese. What surprised me is that the waters did not mix together for many miles.

The Rio Branco brought with it shallow water and much slower progress and by three o'clock in the afternoon we were aground for the third time. All hands jumped into the water to try to shove the creaking vessel off the reef. Plinio seemed to find all this amusing but all I could mutter was why didn't these people, with their years of Amazonian experience, know the river channels better?

Half an hour after a spectacular sunset we beached the boat – purposely this time – on a sandbank. I jumped ashore and, holding my torch, went to investigate the pools and forest edge. There were seven small spectacled caiman around the first three pools, their eyes shining as red as rubies. Two jabiru storks, like Goliath avian sentinels, strutted a few yards to get away from the torch light then, flapping their enormous wings, lifted into the air. They only flew some 150 feet before they landed on the sand again.

I found two skimmers' nests, each with two eggs, and my torch picked out a pair of scarlet ibis on the water edge, probing the wet sand with their long curved beaks. Chris yelled from the boat that dinner was ready and I walked back to a fish feast – a change from cheese and ham! One of the great gastronomic pleasures in Amazonia is the fish, and the varieties are staggering. This night we tucked into the famous tambaqui – a fruit eater that grows to a massive 100 pounds. Justino, belching loudly, offered me the tambaqui head, the best part of a fish to a Caboclo. I politely refused and watched him suck away at the marble-sized eye sockets, and gorge on the rest of the skull, the juices running down his grinning chin.

Early on our fourth day, as I climbed over the bow and dropped into the shallow warm water, I shone my torch back at the boat and caught Marciel's smiling face in its beam. Before I could avert the light Fatima's grin joined his over the edge of the hammock. Rather him than me, I thought.

Before breakfast Chris and I were gazing out across the river when we spotted an unusually violent disturbance in the water. I looked through my binoculars and clearly saw a dolphin. Then another! The two of them were side by side, and every so often twisted and made humping movements.

'Dolphins, and it looks like they're mating,' I said to Chris. Before he could say anything the ship's aerosol-can horn sounded, ruining the moment. Captain Cockroach shouted with gusto from the helm, 'Call the American – dolphin fuckee fuckee!'

IBAMA (Instituto Brasileiro do Meio Ambiente) is the equivalent of the British forestry commission. It has isolated outposts in many parts of Amazonas manned by small teams of armed guards whose task is to protect vulnerable species and their breeding sites. The upper reaches of the Rio Branco are protected by IBAMA because the largest freshwater turtle in the world, or terrapin, returns to its nesting grounds there every year. The tartaruga, as it is called in Portuguese, can grow to an enormous 110 pounds. We knew it by its scientific name, *Podocnemis expansa.* Chris had already told me about the tartarugas and I wanted to see the nesting beaches – a film idea already beginning to germinate in my mind. Who would believe that there are terrapins here weighing 110 pounds?

There was a small temporary shelter, covered by a bright blue tarpaulin, on the river bank 100 yards in front of us. At this time of the year the rivers are low and the climb up the bank more than 30 feet. Four uniformed guards waited at the top of a rickety bush-ladder, and we climbed the ladder to meet the men. Incongruously smart in their green uniforms, they grinned enthusiastically at us. Daniel made the introductions and the guards made great ceremony of our arrival. They insisted on a group photograph being taken, all holding their rifles to look like Rambos. They talked so fast I wondered how I would ever be able to converse in Portuguese. They told us that the IBAMA post further up river had more than 5,000 hatchlings ready for release. Chris said that we had to give them a small box of provisions from our supplies. 'It's the way things are done out here,' he said, words that would come back to haunt me a year later.

We left just before midday and an hour later were firmly aground in shallow water. Plinio and Marciel transferred to a small canoe and paddled in front of the boat, using a pole to try to locate deeper water. After a lunch of fried piranha fish – extremely tasty but fiddly to eat because of the many bones – another old river boat came close by. She was called the *Chiquita* and was going all the way to Boa Vista, near the frontier with Guyana. They had an experienced river pilot, a *prático*, with them and were happy to let us follow their lead. Why didn't we have a *prático* with us I wondered?

We had been following the *Chiquita* for half an hour when Fatima appeared with a battered metal kettle and some glasses. This, I was told by my companions, was a national ritual. *Cachaça* is distilled from sugar cane. There are many different qualities of the drink just as there are with whisky, but I had a sneaking suspicion that this was not a single malt quality! It is mixed with water – very little water – and sugar and fresh limes are mashed into it. There is no doubting its strength. Just one glass and the world seems a lot rosier.

We were interrupted by some shouting from the helm, and Daniel leapt up to discover what was going on. Nosiness getting the better of me I followed him. Cockroach had been drinking his own *cachaça* – neat from the bottle – and was clearly very drunk. Much more worrying was the distance between our bow and the *Chiquita*'s stern – less than three yards. He shouted something, which Chris translated as 'He says he's going to follow her like a tic up a dog's arse!' Daniel and Plinio had to pull Cockroach physically from the wheel as our boat almost collided with the *Chiquita.* At her stern two men waved angrily. Plinio took over steering and the others poured Cockroach into his hammock. Minutes later he fell out of it, banging his head badly, but not badly enough to stop him shouting for the 'American's medicine'! I handed him two Neurofen, knowing it would take a lot more to clear the head he would have later.

On the fifth day, we found ourselves stranded in another shallow stretch of water some 250 miles north of Manaus. At this point the crew from *Chiquita* waved goodbye and continued north. The *expansa* terrapin nesting grounds were two days' and nights' travel away, but the draft of our boat was too deep to allow further travel. Nearby was a small community where we were at last able to hire a *prático* to travel with us in a metal canoe.

He was called Ricardo, and was 23 years old. My cynical turn of mind speculated on what attraction us 'gringos' were to people like Ricardo – we must have signalled more money than he could dream of.

Five of us were to make the journey: Chris, Daniel, Plinio, Ricardo and myself. The metal canoe was only just big enough for us and the 40 gallons of petrol we needed to get there. Daniel hoped that the IBAMA outpost would lend us the petrol for our return.

'What if they haven't got any?' I asked Chris.

'Oh, don't worry, we'll sort something out.'

I'm used to risky adventures but I try to make mine as well researched as possible. This trip was worrying me. I looked at the map, and realised with a few basic calculations that we were going to cross the equator – twice!

Ricardo sat at the front, Plinio drove the boat and the rest of us settled in the middle. Despite his official role as *prático* Ricardo, I believe, would have got lost on a guided tour of a corner shop! We frequently hit sandbars, often at full speed. He clearly didn't know the river. The second night, by torch-light and miracle, we arrived at Aricura. I had made a big mistake and had not brought a hat. My ear lobes and nose were bleeding from exposure to the sun, and my accumulated late nights were beginning to catch up with me.

Antonio, the IBAMA guard, was delighted to have five visitors drop in unannounced, the first humans he had seen in 93 days. In zombie mode, I put my hammock up from one of the rafters of his hut, desperate for sleep. He insisted that I drank a coffee and I could almost feel my teeth loosening in their sockets as I sipped the sweet stuff. Antonio wasn't short of words. Out came all the conversation he had held in for three months – and then out came the dominoes. Brazilians can only play dominoes – a national passion – one way. They slam the pieces as hard as possible onto the table top. At least that's the way it seemed. The dominoes, a crackling radio and Ricardo's snoring prevented me from sleeping for most of the short night.

The next thing I knew it was 4.30 am and Daniel was shaking me awake. The terrapin nesting ground was another two hours north and we had to be there at first light if we were to see the hatchlings being released. At 5.30, the night's darkness ebbed as layers of brightening red and orange light spread from the horizon. This area is so far from any pollution that the sun's coming and going each day is almost always spectacular.

The shallow waters were no problem for our fast canoe, with Antonio guiding us. At each curve in the river, we saw vast areas of sands that had gathered at low water season. Some of them formed steep banks at their edges, more than 15 feet high. Sororoca was one such sandbank, or *tabuleiro* as it is called in Portuguese. Its high flat top was visible from half a mile away. I could also see two spindly poles with small white flags flapping at their tops. These were to advise passing hunters that the place was protected.

We pulled into the edge of the sandbank, its slope rising dramatically 15 or 20 feet above us. Three men were hauling two large mesh cages from the water and through the open top and I could see terrapins, no bigger than a Hob-Nob biscuit, swirling about in their thousands. The men chatted to Chris and Daniel as they tipped the contraptions over to let the creatures spill into the river. Chris translated for me. Each year they keep back some 5,000 hatchlings for two weeks, allowing their shells to harden. This, the team believed, would give them a better chance against predators, such as piranha and caiman. Paulo, a short, fat man, was in charge. He had a permanently sad look on his face. He told Chris to tell me to follow him up onto the top of the *tabuleiro* where I could see the nests themselves.

It was a struggle to get to the top. It reminded me of when I was a child running up and down the sand dunes at Lytham St. Annes in Lancashire, feet slipping faster as I neared the top. Clawing with our hands we reached the flat summit, and there, stretching away to the distant forest, was the nesting ground. It was like a vast desert – but in the rainforest. It was at least one and a half square miles in area and, Paulo told me, was home this year to nearly 5,000 nesting adult *expansa* terrapins. The numbers were mind-boggling. Each female lays more than 100 eggs so over the hatching month an incredible 450,000 babies would dash for the river.

This immense-seeming number is for a good reason. As I scanned the surface of the *tabuleiro,* as far as the eye could see, hundreds of dark shapes huddled and hopped about. Black vultures, nature's undertakers, were here in their hundreds. They arrive in November when the female terrapins come out to lay and stay until February, when the hatchlings appear. Vultures are normally carrion eaters but here they have learnt to take live prey. It is not as bad as it first seems. The majority of baby terrapins do make it safely to the water, but there are other enemies waiting for them there.

Paulo whistled to get my attention. In front of him there was a pit in the sand. It was cone shaped, about three feet deep, with a sunken spot in the centre. Suddenly a tiny head, the size of an olive, poked through. Seconds later the baby terrapin was scrabbling up the slope of the pit, just like I had done to get to its nest site, to begin its run to the river edge. One after another they emerged. I lay prone on the sand with my face at the edge of the pit. I pointed my camera and waited. I was using a macro lens which, when I looked through it, magnified the scene. Grains of sand began to move and the tip of a nose appeared. Two tiny nostrils opened and closed as it breathed, and then with a push its head appeared. After a few seconds it opened its eyes and saw me. I imagined I could almost hear it say 'who the hell are you?' I took a picture of it and then it was off up the slope. I stayed there another half an hour until no more baby terrapins appeared. It was a wonderful experience, and I was already thinking what a fascinating film sequence these terrapins would make.

Before we left, Paulo told Chris that I was welcome to film their project. I smiled at Paulo and shook his hand, but he still looked doleful.

Five blistering hours later we pulled into the mouth of a creek. We stripped off and dived into the cool water, only to be immediately attacked by a swarm of small fish with an awful habit of biting nipples. Daniel tried to run out of the water clutching his backside with one hand and his groin with the other. He shouted, 'Nick, eets the candiru. Pay particular attention to ere and ere,' pointing at the obvious!

'Old wives' tale,' I shouted as I washed my hair, but didn't feel entirely at ease. The fish moved off, pecking at the soap suds that floated away.

Our terrapin interlude over, we eventually arrived back at Cockroach's boat and set sail for the Amazon Association's Xeruini reserve. The River Xeruini was too low for us to take the big boat in, so once again we set off by small canoe to camp for a couple of days further up the creek. But the environment, although pristine, was mainly low riverine forest, with little wildlife to see. My companions kept telling me, 'You need to see it in high water time,' but the fact remained that the area was unsuitable for me to film. Chris and Daniel were disappointed but positive that they could still find me that elusive place to film.

Chapter Nine

Paradise Found

We were soon back on the Rio Negro, heading for a creek some two days' sail away, north of Valdemar's house on the Rio Jauaperi.

'Just what you're looking for, Nick,' Daniel said with authority. I was happy with the suggestion, remembering my night exploring the creeks with Valdemar.

On our way, we called in at the Amazonian equivalent of a supermarket, a floating provision post in a small channel off the main river, at a place called 'Paraná da Floresta' (literally meaning 'channel of the forest').

The raft floor was about 65 feet long by 25 feet wide. Its base was made from four gigantic tree trunks, each at least three feet in diameter. Built on top was a single-storey hut, the shop. The internal walls were shelved and filled with five items: *cachaça,* vodka, sugar, salt and brillo pads. There were no windows – light came from the double door opening or from oil lamps. As I stood in the doorway, uncharacteristically wearing a straw hat that Daniel had given to me, my shadow stretched almost to the counter. I felt I had stepped into a western.

Lying just inside the doorway, to my right, was the biggest terrapin I have ever seen. It was an adult female *expansa,* and she was cruelly laid upside-down on her back. Her head dangled seven or eight inches onto the floor. A pool of fluid was seeping from her mouth.

With Daniel translating, I asked the man behind the counter what it was for. '*Para vende,*' for sale, he replied.

'How much?' I asked. He told us that it was $50. I tried to lift her, testing the weight, but it would have taken Chris, Daniel and myself to have lifted her clear of the floor. She had obviously been caught while she was laying eggs, and to think that this 110-pound creature, probably 30 years or more old, was going to end up on some dinner table in Manaus was extremely upsetting. In the opposite corner was something almost worse, if that were possible. One of the rarest of Amazonian animals, the manatee, a huge aquatic mammal, was represented by a pile of butchered chunks of meat. The man behind the counter knew this rare animal was a protected species but didn't give a damn – and that's the way many things are out here.

While our captain, Cockroach, tried to buy gasoline from the store owner, I went outside. I saw that the riverboat we had tied up to was stacked full of white plastic boxes called monoblocks. Inside each one were hundreds of different exotic fish, many floating on the surface dead or dying. It turned out that the boat owner plied his trade between Barcelos, on the Rio Negro, and Manaus. He was only paid for the fish that arrived alive and to increase their chances of survival he added a dose of chemical to slow down the fishes' metabolism. For every live fish he was paid the paltry sum of ten cents.

There was no gasoline for sale, and we set off for the mouth of the Jauaperi, and Valdemar's house again. The plan was to drop Daniel, Fatima, Marciel and me at an abandoned hut, close to Valdemar's, where we could pass the night. Cockroach would then take the boat down river to a place where he could buy gasoline for our outboard engine. He would return to pick us up the following day.

We waved goodbye to Cockroach and paddled towards our abandoned hut. It turned out to be a ruin. We just about pulled together enough floor boards to keep us all off the bare ground and away from snakes. Fatima and Marciel prepared fresh fish over an open fire and we ate as the blazing colours of an Amazon sunset filled the sky around us. After fixing poles to hang our hammocks, we paddled over to Valdemar's house about half an hour away. As we arrived we saw him canoeing around a corner in the river, being paddled by three of his young daughters. We gave him a large tucanare fish that Marciel had caught earlier in the day, and a chicken that we had brought with us for emergencies. He was more delighted by the chicken.

When we returned to our base, we discovered that the small hill our shelter stood upon was an ant colony. The voracious foragers were all over our provisions and bags. But we kindled the fire and sat around talking about the forest – its darkness and people's fear of it. Fatima was terrified of almost all animals, unless they were dead or being prepared for the cookpot, while Marciel hated snakes and a giant sloth that breathed fire. We climbed into our hammocks at midnight. The sounds of the forest were close enough to touch. Howler monkeys, with their spectre-like screams, and giant rats coughing across the creek to each other kept us all drifting in and out of sleep. Daniel, rifle at his side, kept vigil. He had convinced himself that he had heard a jaguar calling close by. At one point in the early hours he started to explain to me how Brazilian people made love in hammocks. 'Rule number one, Nick, is that position missionaries is not advised.' Daniel's face was side-lit by the moon, he was smiling. 'Also bad for man to kneel – he can fall out of hammock easily. See, Nick, you must have calm and spiritual together, and be corved like furteses!'

The howler monkeys started up again and took his mind off sex.

Daniel and I were already washing in the creek when Valdemar appeared at 6.30. Afterwards we boiled water for coffee and ate Jacobs cream crackers smeared with some dreadful sugary preserve called *doce* in Portuguese – it means sweet, and it certainly is. No wonder gums are predominant in Brazil.

Valdemar took us into the forest along one of his hunting trails. The forest was dark, with less than two per cent of daylight filtering through the overhead trees. He had two small terrier-like dogs with him that were constantly on the go yapping as they chased some poor rodent into a burrow or rotting log. Valdemar told me he felt safer with the dogs in case there was a jaguar about.

Daniel and I stumbled along with Valdemar for four hours. Drenched in our own sweat, we somehow managed to walk into every sharp, sticky, spiked or stinging thing. We were pulled at by thorny vines, snagged by palm tree spines and bitten by an endless variety of ants. We moved off the hunting trail many times following our leader as he cut clear a path with his machete. Unfortunately, he only cut for his own height. Our heads, a good 10 inches above his, caught almost everything that snaked down from above.

It would have been easy to become lost without Valdemar as a guide. I pushed that thought from my mind as he showed me the tracks of a tapir in the soft mud on the edge of a creek. '*Muita comida*,' he said to us. Daniel translated – a lot of food!

I must have fallen through the rotten floor, where large trees had once stood, at least four times. Up to my waist on one occasion, in a dark wet pit, I struggled to pull myself out and away from the poisonous snakes I imagined lurked below. The ants were worse. They showered us as we bounced off the vegetation. They went down our necks, up our trouser legs, and into our hair. Finally, weak from the battles, I leant against a tree trunk to mop the sweat from my face. Valdemar said something to Daniel and I recognised the Portuguese word for ants – *formiga.* I interrupted to say that I was sick of the damned beasts. As I finished berating them, I felt a sudden swingeing pain in my hand. My frantic reaction to it had Valdemar almost on his knees laughing. He had been trying to tell me, through Daniel, that the tree trunk had a nest of ants on it. They were called *jaguar ants!*

On our way back, although Daniel and I were not aware we were heading home, Valdemar showed me an ants' nest and said it would bring rain if he set fire to it, and a tree whose wood was good for fashioning paddles, and its bark for a medicinal brew. It apparently cured sickness of the 'head, liver and bottom'! Daniel translated as Valdemar also informed me that the local howler monkeys only defecated from one tree and that when you shoot one, the rest of the group 'sheet themselves and you get covered in the steenking stuff because they have no anal muscles.'

'Getting their own back,' I said.

Daniel and I leaned against the rickety frame of the ruined hut feeling exhausted, while Valdemar still looked as though he could run a marathon. Cockroach and his boat had returned, with gasoline for the fast boat, and we said goodbye to Valdemar as we left to rejoin the main river channel.

As the next day dawned, I was roused from my hammock by Cockroach screaming from the helm, 'Call the American, call the American!'

The monstrous head of a black caiman was gliding across the surface of the river about 100 yards in front of us. We watched this prehistoric beast swim along the surface, and then slip beneath it, as our boat closed in on it. Cockroach told Chris that the black caiman's penis was good for men to attract women. It is ground into a powder to make a paste. The man then paints the paste onto his own member and, so the folklore goes, the next women he approaches simply cannot resist him. I told Daniel to tell Cockroach that in Europe we use oysters as an aphrodisiac. Cockroach, grinning like a pervert, asked Daniel if we prepared oyster paste in the same way!

By eight o'clock, true to form, our tub was grounded in shallow water. Further progress up the Jauaperi was impossible. While we were standing in the river trying to push the *Barata* into deeper water, Daniel shouted, 'Nick, you have a *sanguessuga* on your arm!' The black slimy creature clamped onto my upper left arm was clearly a leech. It's funny how the mind works.

It didn't bother me one bit, except for feeling indignant at its very existence – I had read on the very first page of Redmond O'Hanlon's *In Trouble Again* that there were no leeches in the Amazon. I felt like taking it home.

It took us six hours by canoe to reach the Xixuau creek. Justino, wearing a hat that made him look like a giant garden gnome, and Plinio, came with us. Many times we had to pull the heavy metal canoe with its outboard engine across large expanses of sand covered by only six inches of water. The dry season was taking hold. The water of the Igarapé Xixuau is crystal clear and Daniel, Chris and I were mesmerised by hundreds of fish darting to and fro: schools of pacu, tucanare, aruwana, pescada, sardinha and piranha. It was a remarkable sight.

With high forest on both sides of the creek, we followed its course as it twisted and turned for about a mile and a quarter. Justino cast a hook and line over the side and within seconds pulled a huge tucanare fish on board. It must have weighed at least 12 pounds. It was golden yellow and black and had an enormous bulging forehead. Justino said this bulge meant it was a male. He caught another two within ten minutes. It was beginning to get dark so we turned back to find the Caboclo whose home we were going to stay in for the night. Our dinner was lying gasping in the bottom of the canoe. I asked Chris to get Justino to put them out of their misery and he bashed them with the machete.

Carlitos and his family live on the edge of a small lake near the mouth of the Xixuau creek. With his wife and six children he survives by hunting and fishing. Like his father he was born here. His closest neighbours were a day's paddle away. We were the first people outside his family he had seen in more than two months. He remembered Chris and Daniel from their previous visit and welcomed the visitors, the obligatory food parcel, with tobacco and *cachaça* catching his eye.

Carlitos was about 40, very thin and walked in what I can only describe as a floppy sort of way, his arms dangling loosely by his side as he walked. He had a bad squint and this somehow gave him a look of being drunk. He was shy but smiled easily. His wife was quite different. She looked almost pure Indian, and wore a fierce expression. We never saw her smile. As it turned out, she had good reason for being like this, because her husband's other main quality was laziness and she was left with most of the hard work.

The immediate area around Carlitos's house was filthy. Bits of rusting tin and plastic littered the ground. The bare earth next to the back door stank of urine. Fifty yards behind the house was a cleared patch of land with ten small wooden crosses sticking up from the earth. This was the family graveyard. Five of Carlitos's other children were buried there, alongside uncles, aunts and grandparents. It was a stark reminder of the harsh environment in which they live.

The steep bank leading down to the water's edge produced a surprise, a stone axe head which proved that the Indians had lived on this site a long time ago – research showed it to be 2,000 years old. No doubt the Indians had settled here for the same reasons as Carlitos's family – the rich waters and easy fishing.

As a young boy, Carlitos was initiated by his father into the ways of hunting in the forest and creek. During the 1950s and 1960s the world was hungry for the skins of giant otters, jaguars and manatees, and the trade in these pelts reached even the most isolated parts of Amazonia. Carlitos and his family lived in part by hunting these animals. Then, some 20 years ago, the world changed its view and the trade in these now endangered animals was outlawed. After eating the delicious tucanare his wife had cooked, we slung our hammocks from the fragile looking timbers of a room at the far end of the house where the family normally fed and relaxed. The thatched roof had many holes in it and Carlitos told Daniel that his wife was 'on him a lot to fix it'. Carlitos squatted on the floor and eagerly swigged from a bottle of the *cachaça* we had given him. Within half an hour he was plastered and the expected jungle tales surfaced. He had seen 26 jaguars, and various monsters that no other man had seen.

At one point he reached for an empty two-litre plastic Coke bottle we had brought with us. He cut a part of the neck off, the knife slipping and only just missing his thumb, and then blew into it to make the most horrendous growling noises. Eventually, with glares from his wife, he put the bottle down and told Daniel and Chris he was calling for jaguars. This, he said, was what you had to do if you wanted one to come to you.

The following morning we were awakened by a group of howler monkeys on the other side of the lake, swinging between the tree tops in the early warm sunlight. We drank cups of sickly sweet coffee and chewed on some dried biscuits before getting into our canoes.

I was in one canoe with Justino, while Chris and Daniel were in another. We paddled upstream for two hours at which point Daniel and Chris stopped for a swim, while Justino and I carried on alone. Fifteen minutes later we rounded a bend to find ourselves surrounded by seven giant otters. They huffed and puffed only a few feet from us, bobbing up and down treading water. They showed no fear of us – another cue for my camera!

Later Justino and I were suddenly confronted by a brightly-banded red and black false coral snake – so called as they mimic the colour of their very poisonous relative, the true coral snakes. I remembered the rhyme my supervising producer at Survival, Mike Linley, an expert herpetologist, had recited to me before I left England:

'Red on black, venom lack
red on yellow, kill a fellow!'

With that little ditty firmly in my mind I signalled to Justino to stop. I pointed at the snake and mimed that I wanted to get it. Justino looked horrified. I took the paddle from him and pulled us over to the swathe of brilliant green aquatic grass that the snake was resting on. It was about two feet long and strikingly beautiful, especially against the green background. I tried to get the paddle under it to bring it on board but as I lifted it clear of the water Justino lurched, tipping the canoe violently, and the snake slipped off into the water. Justino waved his finger madly in a 'no' gesture at me. I signalled to reassure him.

The Xixuau creek is a naturalist's paradise at this time of the year. Toucans, macaws, parrots, hawks, ibis, oropendolas and three species of kingfisher were common sightings. On this stretch alone we saw 11 giant otters, a sloth, a tamandua – a tree-dwelling ant eater – and an anaconda. On the last stretch before we entered the lake where Carlitos's house was, the water in front of our canoe heaved upwards as though a small submarine were about to surface. This was caused by one of the rarest creatures in Amazonia, the manatee. It is a huge doleful looking animal with a large paddle-like tail. It swims slowly most of the time, grazing on aquatic vegetation. It has been hunted for many years both for its meat and its extremely thick skin, which was used in developed countries for machine belts. Justino cupped and opened his hand a few times in front of his mouth to signal 'good eating'.

Back at Carlitos's house, Chris and Daniel could tell from my excitement that I had found an area to film at last. Over dinner of fish and rice we talked about the logistics of setting up the film. I explained that all the projects had to be cleared with Survival first, but that it looked promising. Chris then told me that the Amazon Association had decided to have their reserve here, in the Xixuau, and that Carlitos had agreed to sell his land to them.

Looking at Carlitos, a drunk in his hammock, I wondered how he could have concluded anything in his present state. I thought his wife would probably kill him when she found out.

The following morning I bathed and shaved in the river. There's little to compare with the clean feeling a wet shave gives you when you are living in these surroundings. Three of Carlitos's youngest children played noisily in the water close to me. Afterwards, Justino and I set off for our final day's exploration in Carlitos's small dugout canoe. As we paddled further upstream I looked at my tatty map of northern Amazonia and realised that, tracing the route of the Xixuau, no other person lived between where we were and almost 120 miles away. We saw several river turtles, one of them the *expansa,* the same species as nested on the Rio Branco *tabuleiros.* I resolved that if further research showed that these giant terrapins migrated to those nesting grounds, we would come back and film the story.

Before we could leave the Xixuau, I wanted a last walk through the forest close by Carlitos's house. We were up at dawn and after the usual coffee set off with Carlitos and Justino along a well-worn hunting trail. Half a mile from his house the forest was pristine, untouched and wonderful to walk through. Enormous trees, some probably around 200 years old, towered above us, and only tiny darts of light penetrated the almost continuous canopy. The floor was soft underfoot from the thick layer of decaying leaf litter and the air was heavy with the smell of decomposing plant material.

Just as we were about to leave the creek and head back to Cockroach's boat, six hours south, a large wooden canoe pulled into the lake. There were five Caboclo men in it, two at the front and two at the back, with one obese man sitting in the middle, his arms folded over his enormous belly like an oversize Buddha. There were also two mounds covered with blue tarpaulins. Lying on top of one was a big dead sloth.

Carlitos walked down to the edge of the lake to talk to them. They were traders and, lifting the cover of the mound without the sloth on it, tried to sell Carlitos some turtle meat and bottles of *cachaça.* They pulled out a shotgun and showed it to Carlitos, but he wasn't interested. It is at moments like this that I feel so powerless. All the ecological regulations in the world seem to pale into insignificance when faced with the reality of people desperate to earn a living.

The voyage back to Manaus was fittingly eventful. Cockroach's tub broke down, mid-Jauaperi river, just as I was relaxing in a shower. I had a sneaking feeling that Manaus was further away than I thought. As I rinsed my hair with a bucket of water, I heard Chris shout that we had hit rocks and the propeller shaft had sheered off. Justino went off in one of the dugout canoes to find help. I hadn't seen another soul since we left the Xixuau and wondered where on earth he would find help. But luckily, as I found out later, he used to live in the lower reaches of the Jauaperi and knew that about two hours' paddle away was another settlement. To the surprise of all of us, he returned with a much smaller boat, but one with a working engine. She was 20 feet long, single-decked, and falling to bits. This was the boat that was going to tow us for the next two days back to Manaus.

Chapter Ten

Caught in a Coup

Manaus was a welcome sight after three weeks up river. The first thing Chris and I did was to jump into a taxi and speed to the Anaconda Hotel. As I waited in reception I was gripped by the manic need to find at least one uncracked ceramic floor tile! But at least my room had a bathroom and I ripped off my dirty clothes and spent half an hour indulging in a hot shower.

The following two days were hectic, running around various scientific institutions, amongst other things. Everyone seemed happy to help should Survival decide to make films here. I discovered that inflation was running at 40 per cent per month, so working out a budget was going to be extremely difficult. With my bags packed I said goodbye to Daniel and Chris and promised to let them know the moment a decision was made. My flight back to Caracas was at three in the morning, leaving a few hours to explore Manaus.

The streets were untidy, extremely busy and noisy. It took me a whole ridiculous hour in the Banco do Brasil to change a $50 traveller's cheque, and the second I walked out a street vendor stopped me. 'You will speak German,' he demanded.

'No, I am English,' I replied.

'Ah, then this one is for you,' he said as he delved into the plastic bag hanging from his left wrist. He produced a small plastic wallet with a Union Jack on it and he squeezed it. God save the Queen tinkled out cheaply…ding ding diiing ding ding ding…

'What is it?' I asked.

'Very very cheap for you, my friend.'

'Fine, but what is it?' I insisted.

'Musical *camisinha.*' *Camisinha* in Portuguese means little shirt, and that is the affectionate term for what we call condoms. I made my excuses and left, as they say.

At the airport, I was sitting in the departure lounge making small-talk with a friendly bartender, when he told me that there had been a coup in Venezuela five days ago! I almost choked on my Diet-Coke. He'd seen the reports on television – people were shooting each other in the streets of Caracas. No one could give me any further details. I was desperate to know if Gordon, Jorge, and Charlie were safe and immediately telephoned Mike Linley in England, apologising for waking him up. He had heard the news himself and had called the Venezuelan Embassy in London for information, only to hear that all lines between Caracas and the outside world were closed.

Mike wished me luck, told me to be careful and then remembered to ask me about the recce in Brazil. 'By the way, Nick,' he added, 'you didn't come across any corals, did you? Because I have since found out that the false coral snakes in that area where you were don't follow the rhyme rule!' I could see the panic on Justino's face, his stubby finger waving at me like a crazy metronome.

The flight came in and my mind eased a little when a steward gave me a clearer picture of the situation in Venezuela. It was a military coup, and mainly centred around Caracas. As far as he knew the southern wilds were unaffected. The airport had been closed, but had re-opened the day before. What timing, I thought to myself.

I passed through immigration and tried to get into the domestic wing for a connection to Puerto Ayacucho, only to find the large sliding glass doors barred by a line of seven armed soldiers. The atmosphere was tense as I returned to the main hall. There I learned that no internal flights were operational until further notice. In desperation I called Charlie Brewer.

'Hi, Nick,' and as though nothing had happened, 'how was Brazil?'

'Better than here by the sound of things, Charlie.'

'Yeah, things have got a little hot, Nick. Better we don't speak about it on the phone if you understand what I mean?'

'OK Charlie, just tell me what you know about the chances of getting to Ayacucho – Gordon's still there.'

'Don't worry, Nick, Ayacucho is unaffected and the airways will be operating from tomorrow morning.'

I spent that day and the following night shut in the airport. No one was allowed out of its precincts to go *anywhere*! The place resembled a refugee zone as thousands of stranded passengers jockeyed for a space on the floor. There was no food, no drink and – worst of all – no functioning toilets.

By dawn the message had filtered through that internal flights were running normally. The one to Ayacucho was third on the list.

I arrived to find Gordon and the gang in Ayacucho alive and well. They had missed the coup. Gordon actually felt a little disappointed as the only visible signs that something was untoward were military roadblocks around the town. The town was a little short of supplies, but Julio's pool bar had plenty of beer, so their world was at peace.

I told Gordon all about my trip and my hopes that we would return together to make our next Survival film. He liked the sound of that, but for now we were both eager to finish the tarantula film and get back to the UK for a break. Family, home, toast and marmalade, cups of tea and a cool climate were less than three weeks away.

The last few days were spent tidying up the last sequences in the film sets and shooting most of the close-up scenes of *Theraphosa*'s anatomy. Winding down is always an extremely pleasant feeling, and I was looking forward to getting involved in post-production work – editing, sound tracks and commentary all had to be done. Having shot the film over such a long period, seeing the show come together on a film editor's bench is strangely exciting. Sometimes, though, I would feel extremely out of place trying to readjust to the culture of my own country, my mind still in the forest walking the trails, picking up snakes and overpowered by the atmosphere and isolation.

After a 10-hour journey from Caracas to Heathrow, I was back in the UK. After only half an hour in Britain, I already felt claustrophobic among the crowds of passengers scurrying along the corridors of the airport. I telephoned Emma from the baggage reclaim hall and she whooped with joy when I told her where I was. My bags and equipment officially cleared, I pushed two trolleys in tandem, and scanned the dozens of people waiting for people emerging from Customs.

Many people held placards with odd sounding names. Even when I am not being met at the airport my eyes are compulsively drawn to these rows of placards. As usual I read them avidly. Mrs Colgate from Toronto, Marconi, B. Burruti from Tanzania, W. Anker from Venezuela!... Mike Linley's smiling face appeared from behind the card.

He drove me out of the airport complex onto the M4, and I opened the window to take deep breaths of fresh British air. Even the motorway traffic looked good. Jimmy Young and Ken Bruce were still on Radio Two. It was wonderful to be back. After a few days in Norwich sorting out post-production schedules, it was time to go home and relax. I first wanted to visit my mother in Lytham St Annes, then my osteopath, and, finally, to get on the ferry to the Isle of Mull. There I would go into my house, open all the windows, step out onto the balcony and, with Emma sitting on my knee, breathe in the view of the Sound of Mull.

A few days at my mum's, being pampered, nipping out for – what else? – fish, chips and mushy peas, not having to think about *anything* except the next cup of tea or hot bath, is the perfect way to relax. Frances, the family housekeeper for the past 20 years, imparts the wonderful, comfortable feeling that despite my having been away for a long time, the world has not changed. She makes us laugh out loud with her own brand of Spoonerisms. After her welcome-home hug she pottered through to the kitchen to put the kettle on, exclaiming 'I don't know how you can film those horrible beasts – they make my blood curl!'

My back had been playing up all through the tarantula project. Lifting heavy gear and stooping over cameras in contorted positions for hours at a stretch over nearly a decade has taken its toll. Luckily I found someone who could cure those ills some years ago. My osteopath, Keith Brimelow, lives in Oban and that is where I was heading on 28 May, 1992. The five-hour drive north would normally have been a pleasurable experience. However, by the time I arrived in Oban my back had seized up and when I tried to swing my legs out of the car door I yelped with pain. I managed to reach for the carphone to dial Keith's number.

'Hello Nick, how are you? Back from your travels?' I wanted to laugh at the unintended pun but it would have hurt.

'Actually, Keith, I am outside your office now,' I gasped. 'Can you please come and get me, I don't think I can walk.' Together with his receptionist he eased me out of my car and we waddled into the surgery. Fifteen minutes later I could have done a fox-trot! I laughed with the relief and we chatted about my recent adventures. I told him he would always be welcome to travel with me if the idea ever appealed. It didn't.

My house was still standing, the garden looking lovely, the weather a perfect Scottish spring day. Inside on the kitchen table I looked at a pile of eight months' mail – and decided to leave it until later. I rushed to call Emma, who lived with her mother under a mile away. The machine was flashing telling me there were over 100 messages – they could wait, too. Emma had just got back from school and came straight around. 'Haven't you looked on the kitchen wall yet, Daddy?' There in all its innocent glory was a water-colour painted poster that proclaimed 'WELCOME HOME DADDY'. My eyes were wet with tears as I gave her a huge hug. Only three more films, I said to myself, then home for good.

Fifteen weeks flew by – it felt as though I had hardly unpacked before we were booking flights to return to South America. To my relief, Survival was keen on the idea of me making several films in Brazilian Amazonia and commissioned the first film, *Creatures of the Magic Water,* to be shot in the areas I had recently visited. My contract was for 18 months, the story line agreed. The logistics for the new project were quite awesome. We were going to be working in two locations hundreds of miles north of the nearest city, Manaus. The locations were a five-day sail away from civilisation so we needed to buy a good riverboat to ship ourselves, 40 equipment cases and seven tons of scaffolding (to build canopy filming towers), from Manaus to the middle of northern Amazonia. Everything from work permits to bank accounts, contingencies for emergencies, to the minefields of Brazilian bureaucracy had to be sorted out before we set off.

My penultimate day in the United Kingdom was typical. As time starts to run out, things always go wrong. I was at the counter of a bank in Preston asking for the £30,000-worth of American dollar traveller's cheques. Unfortunately their office in London had misread the order and only sent $30,000. It took four hours to sort out, helped a little by the steam coming from my ears and nostrils, and then the hand-cramping task of signing every individual cheque. The clerk and I laughed when, during the fiasco, a drunken vagrant in the bank's doorway burst into loud song, *Que sera sera.* I am sure I was laughing for a different reason than the clerk.

It was a relief to reach Manaus, which Gordon and I did on 28 September, 1992.

Chapter Eleven

Ma-Nauseous – the Great Rip-Off

The circuitous route to Manaus, Brazil's jungle city, took us through the airport of Rio de Janeiro. Like zombies with a bad case of body odour, we gathered all the equipment from the baggage hall and were welcomed with open arms – by the Brazilian Customs men. They couldn't speak English, and we couldn't speak Portuguese. Deadlock. What was obvious was that they would not give us permission to enter Brazil with our gear.

Our permits and visas had been organised through the Brazilian Embassy in London, and for once I really had thought everything was in order.

This clearly was not the case and, it turned out, their officials in London had given us the wrong forms. At six in the morning a woman was brought in who could speak a little English – 'good evening' – and she told me that we would have to stay in Rio for the next four days while they processed our application to bring the equipment into the country. 'Impossible,' I said in a deliberately too loud voice. Gordon winced with embarrassment.

It was too early in the morning to speak to anyone, except Survival, which was five hours ahead of us. I telephoned Mike Linley in the UK and he immediately raced into action, with a flurry of phonecalls: the British Embassy in Rio, Chris Clarke from the Amazon Association who was waiting to meet us in Manaus – and the Brazilian Consulate in London. The clock ticked away as the time to catch our connecting flight to Manaus came closer. Gordon and I sat down on a broken luggage conveyor belt, with hangdog expressions.

'Why can't it ever be simple?' I asked him, remembering China, Qatar, Israel, Guyana, Sierra Leone, and Venezuela.

A cockroach scrambled out from the conveyor's rubber rim and Gordon stamped on it. 'Here we are again,' he said.

It was the Rio Branco river turtles that saved the day. Chris Clarke managed to get hold of the superintendent of IBAMA in the State of Roraima (the man who ran the turtle project) and persuaded him to fax the authorities in Rio saying that Survival's team just had to get to Manaus '*agora*', now, this very day, with all their film equipment because the pregnant turtles were already coming out to lay eggs at their site.

The burly Customs man was not happy, but an obviously higher authority had told him to let us through. We smiled smugly as a parting shot – a big mistake. As a final act of nastiness he altered our equipment permit from two years to one – which was to have knock-on effects much later in the year. But for now, with less than one hour to make our connection, we were legally in Brazil.

Chris and Plinio met us at Manaus airport with a knackered-looking three-ton open-back truck. With all the film equipment loaded, we set off into town to find our temporary accommodation. Gordon, always understanding my paranoia, offered to travel on the back of the truck with the gear – bless him.

The Janelas Verdes (Green Windows) Hotel, with its distinctive yellow painted windows, was near the port. We had stayed in better establishments over the years, and worse! We piled our equipment and ourselves into two small rooms smelling of Gorgonzola cheese and damp cupboards. Chris was in a flap to get me down to the port to look at the riverboat he and his fellow Amazon Association partner, Daniel Garibotti, had found and thought ideal for our project. Survival had agreed to donate the boat to their organisation at the end of the second film project, three years down the line.

The *Commandante Brandão* was enormous. Seventy tons of wood and a Caterpillar diesel engine. She had two decks with six cabins, an 80-foot square saloon on the upper level, with a bar (my goodness, what will Survival say?) opening out onto a large sun-deck. On the lower deck was a dining area, galley, bathroom and engine room. Apart from thinking she was too big, I noticed that she had obviously just been repainted – to hide what, I wondered? Chris, detecting my hesitation, said that second-hand boats of this quality were as rare as hen's teeth at that time, and that their values were going up by the day because of inflation. They had taken her out for a test run. 'She is perfect,' he told me.

I had to trust their judgement – after all the association had eight years of experience travelling the Amazon region – and handed over $21,000 to a very happy ex-boat owner. Watching him count the money I couldn't help feeling that I was making a big mistake.

The following morning we discovered three things. First, overnight in a far away land called England, the stock markets of London had crashed, effectively wiping £24,000 from my film budget. A quick call to Survival brought relief from that worry as they put wheels in motion to approve compensation for the loss. Second, we saw that a veritable crowd of people were gathering outside the boat to wait for 'Survival' to set sail for the northern waters of Amazonia.

First was the crew – including the Amazon Association president, Plinio, as captain; Benedito, a fat mechanic with dodgy breath; Manuel, one of Carlitos's sons, as auxiliary mate; and Justino's tired-looking wife as cook. In addition there was Eric Falk, a friendly Dane with a good sense of humour, the third partner in the Amazon Association; João, the jungle expert; Justino, who had accompanied me on my reconnaissance; Justino's three grandchildren; and two friends of Chris's – Alan, a hospital technician from England, and Greg, a youth counsellor from New Zealand. At that moment I said sarcastically to Gordon, 'I wonder who's being taken for a ride?'

And finally, we discovered the *Commandante Brandão* needed a huge amount of work before we could set off, not least complete electrical rewiring.

'No problem,' said Chris. 'Alan is an electrician.'

Alan was a charming man in his late 40s, at times with a terrible stutter. He had come to the Amazon to get away from some horrors back home and to sate his passion for photography. He happily agreed to re-wire the boat for nothing, as we were, after all, giving him a free trip to the Amazon Association's new reserve, the Xixuau.

Alan had arrived in Manaus that morning. He had foolishly climbed into an unregistered taxi at the airport and shown the driver a piece of paper with the hotel's name on it. Twenty minutes later the driver had stopped, pulled a knife on him, and demanded '*dollares*'! Alan handed over his wallet, which contained all his money – pounds, dollars and traveller's cheques – his return flight ticket and his passport. The robber got back into his car, drove a few yards, then reversed, throwing something at Alan. He picked the bundle up – it contained his sterling currency. Perhaps the thief had been reading the *F.T.*!

With the costly boat repairs underway, Gordon set about buying provisions for the first expedition to the turtle nesting grounds. The turtles had already arrived and we needed to get to the site quickly. Our seven tons of scaffolding for building filming platforms was still somewhere mid-Atlantic, so I decided the best way forward was for me to go alone on the first leg, while Gordon stayed in Manaus. Here he could find an apartment to rent, buy furniture and, most important of all, clear the scaffolding with the customs when it arrived. I would return in four weeks' time to pick him up and take him to the second filming location.

At the end of the third day's madness, Gordon and I flopped onto our beds and opened two bottles of ice-cold beer. 'Manaus,' I exclaimed sarcastically, 'I quite like it, despite the dirt and noise.' There's a suburb not far from the National Amazonian Research Institute, called Tira Dentes – it means 'pull teeth!' Great address, eh? Down the road from this hotel there's another called The Cheep Hotel. It's got a balcony over the seedy doorway with three prostitutes on it touting for business. I think we got the best end of the bargain with this place.

'How did it go with Chris?' I asked Gordon, who had been rushing around all day buying supplies.

'Good job we didn't let him do it alone – he thinks you're a pot of gold. At one place he said to me, "we need this". It was a bloody pack of 24 toothbrushes! I told him no way, Survival won't pay for those things,' Chris said "but it's a bugger when you lose your toothbrush in the jungle!"'

This incident summed up an increasing problem for us. I was on a strictly fixed budget and had seriously under-estimated costs for this project. Somehow I had to control it or major problems lay ahead.

On our fifth night in Manaus, once again exhausted from the dreadful humid temperatures, we invited Alan to our room for a beer. This was also the night Chris had asked some people around to be interviewed for a possible job with us. We needed a person who could organise the boat and crew, help with some logistics – and help us learn Portuguese! After this first trip Gordon and I would be on our own working among Portuguese speakers.

Alan was just telling us the most tragic tale about his ex-wife's death when there was a knock on our door. Chris entered with the most stunning young Brazilian girl. He introduced her, 'Nick, this is Antonieta, she's interested in working for you.' She put her hand out and I shook it, mesmerised by her face. She just said 'Hi'. Chris told me to come across to his room when I was ready to interview her and the three others who were waiting. They left our room and we all turned to look at each other, speechless. After a second or two we all burst out laughing and Alan stuttered, 'I be-bet sh-she-she'd be goo-goo-good for the co-coo-cooking alone.'

'You can't go up river with her unless I'm with you, you rat.'

Gordon's reaction was understandable, I told him, but added, 'think of it as a sacrifice – she'll still be here when I get back. You have important work to do, like buying furniture. Anyway, chaps, I must go and interview her now, if you will excuse me.'

'She's fu-fu-far too yu-yu-young for you, Nick,' stuttered Alan.

'What about Michael Caine?' I retorted. Sniggering, I left them in the room to find Chris and Antonieta, hoping that the other three hopefuls would be ugly old slappers.

Antonieta was 19, well 'nearly 20' I told Alan later, trying to minimise the difference. She had got the job. 'Her credentials are perfect, Alan, she's learning English, she is a legal secretary,' and then, lost for words, Gordon offered, 'She's got a great body!'

'Exactly,' I said. 'Anyway her boyfriend was there, he's an artist, and he looks double her age. She is obviously attracted to older *intellectual* men.'

'That counts you out then,' quipped Gordon.

The day of our departure to the terrapin nesting grounds dawned. I was up early and went to visit the National Amazonian Research Institute. I met Fernando Cesar W. Rosas, a scientist working with giant otters and Amazonian manatees, to discuss a future expedition to film river dolphins. I had first met Fernando during the recce and he was delighted that the film project had been given the go-ahead by Survival. I told him that my assistant Gordon was running around Manaus like a kleptomaniac provisioning the boat for the first expedition, and that we were leaving that day for the Rio Branco. Fernando was from the south of Brazil, and didn't much care for Manaus. 'Ah, Ma-nauseous,' he said, knowing much better than I did at that moment what the place held in store for us over the next few years.

At midnight on 5 October, 1992, the *Brandão,* loaded to the gunwales with provisions, fuel and people, finally pulled away from the bustling dockside. Gordon waved to me, looking a little forlorn. I didn't feel too sorry for him as Antonieta had come up with a pleasant solution. She had a girlfriend, Daniella, who could speak some English, and had agreed to help Gordon shop for an apartment and furniture.

Our journey north, to the nesting grounds of *Podocnemis expansa*, the giant river turtle, took five days. The river level was higher than during my visit in February, allowing us to make faster progress, but first we had to call into the Jauaperi and drop Alan and Greg off at the Xixuau.

As we approached the turtle's nesting beach on the sandbanks at Sororoca, I noticed hundreds of small stumps sticking out of the water surface close to its edge. The heads of the adults were gliding to and fro waiting for the right moment to come out onto the edge of the beach to bask in the sun. This behaviour is what is known as 'thermoregulatory' – i.e. it helps regulate their body temperature – and possibly also linked to the internal development of their eggs. The IBAMA guard Paulo was still there and seemed pleased that we had returned, although he looked as glum as ever.

The first thing I wanted to film was the terrapins' 'sunbathing' behaviour, and after that the adults coming out at night to dig their three-feet-deep nests and lay eggs. Walking the sandbanks highlighted the first problem as they were incredibly shy animals. I would have to set up a canvas filming hide and, what's more, get into it before sun came up so that they couldn't see me. The daytime temperatures were too much to bear for any length of time, but in my hide I would at least have some shade.

The canoe dropped me off before first light and I set up the camera on its tripod and waited, and waited. Unbelievably, I had forgotten about the *piums.* At dawn these flies were there with a vengeance. The irritation was almost unbearable. I had a head net to keep them off my face, but that made me even hotter. The hide acted like an oven, and sweat poured off me. In the event I stripped off all my clothes and sat there naked.

When the terrapins started to emerge onto the edge of the sandbank, they did so excruciatingly slowly. I filmed a little but soon realised I would have to move the hide the next day to get a better view. As we pulled away in the fast canoe that afternoon, I saw about a hundred adults basking, necks extended, just out of sight from where I had been positioned.

I spent two further days in the solitary confinement of my hide, until I finally managed to get some pretty scenes of the animals moving out of the water, basking, and returning to the river to cool off. Paulo, with Daniel translating, told me that the terrapins come to the beaches in October every year, and spend the first four weeks in the water, coming out several times each day to bask in the sun. In November, they start to emerge through the night to lay their eggs. The nesting takes place over a month or so, and the adults remain in the water close to the nest site until the hatchlings appear in late January and February. Then they disappear.

This information, together with many other details about clutch size and the like, was useful, but it also made it clear that we had come to Sororoca on a wasted trip. I wasn't going to get the film of the turtles emerging to dig their nests until November. I was furious at having been misinformed. This jaunt had cost thousands of dollars in fuel and provisions and salaries and we still had to come and do it all over again in November. It made me resolve in future to triple-check information from even the most reliable source. What made it worse was that no one seemed to understand how serious it was from a financial point of view. 'What's the matter? You can come back in a month and get it all,' Chris casually remarked.

At least Gordon was pleased to see us back. The good news was that he had found a convenient apartment close to the town centre and had furnished it. We had a home. The bad news was that the scaffolding, generator and diving compressor had arrived, but that customs were not releasing them. This was the start of another frustrating two weeks.

Having worked in African countries, the Middle East, Guyana, Venezuela and China, I am well used to having to face up and deal with corruption. The Third World only continues to work by its own methods. But, in Manaus, we met a whole new repertoire. We were being ripped off left, right and centre, and there was precious little I could do about it. The shippers in London had paid all clearance costs, but that meant absolutely nothing to the officials in the *Alfandega* – Brazilian Customs.

For our equipment to clear the docks we needed a *despachante*, or clearance agent. We were put in touch with Portuguarra by a friend of a friend of Chris's. He seemed OK, if only because I believed, naïvely, that this character was going to be the 'abracadabra' to the customs. I asked Gordon to stay with him as he ferreted about in the warrens of Manaus Customs. Portuguarra had style.

'He seems to be meeting a lot of people,' Gordon commented, 'and he took me for a fabulous lunch out by the airport today.'

'When will the scaffold be in our hands?' I asked.

'He's coming round tomorrow to give us the latest news,' Gordon replied.

A male golden-white tassle-eared marmoset (Callithrix chrysoleuca) *with its 3-week-old twins.*

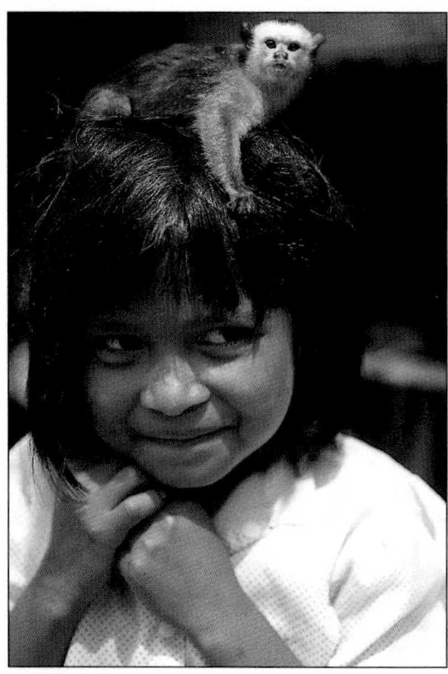

ABOVE *A satarɇ marmoset* (Callithrix satarɇ) *– the Satarɇ women wear them in their hair to eat their nits.*

RIGHT *Things don't always go according to plan – a painful attack on my earlobe!*

The very first glimpse of a pair of satarɇ marmosets – a new breed of marmoset.

A female spider monkey (Ateles paniscus paniscus) *with her baby.*

RIGHT *A pygmy marmoset* (Cebuella pygmaea), *the smallest monkey in the world. The Tikuna women wear them in their hair.*

BELOW *A paca* (Agouti paca) – *a large nocturnal rodent that looks like an enormous guinea pig.*

ABOVE *The ant glove – this painful initiation ritual involved putting a hand in the glove and enduring the onslaught of ants.*

LEFT *Topping off the lethal glove with macaw feathers*

BELOW *Paddling up the creek with my spider monkey, Pete. He always jumps into the canoe for a cuddle!*

INSET *Hauling my camera up the filming tower again*

My two worlds.
RIGHT: *Jacaré creek in front of my house, with a filming tower in the background. The river in the foreground is the only way out – two hours canoe ride to the nearest town.*
ABOVE: *Tobermory harbour – my other world – and where Emma goes to school.*

My daughter Emma. This was taken when she was nine years old, on her second visit to my camp.

The stairs up to my house – Antonieta plays with our woolly monkey Lucy and her 4-week-old baby.

A view of my house from the creek.

ABOVE *The Sataré community where we discovered the new breed of marmoset.*

RIGHT *The rainforest is full of surprises at every turn – just outside my camp, a flowering bromeliad grows through the leaf-litter.*

BELOW *Never let the rainforest lull you into a false sense of security . . . lurking in the leaf-litter is a scorpion female carrying her one-day-old young on her back.*

Pete. I raised him from a youngster. He was my companion for more than three years until he disappeared on 8 February 1997.

Tomorrow turned out to be ten days later, and the man had the brass-neck to sit on my new sofa and tell me, 'Yes, the scaffolding will be in your hands this afternoon.' Antonieta translated, looking worried. In those first few weeks I had picked up a lot of Portuguese words, and especially got to grips with numbers where money was concerned. As our *despachante* talked to Antonieta I heard him say, in Portuguese, 21 million cruzeiros ($2,500). I leapt up like a madman. Portuguarra looked shocked by my reaction.

Through Antonieta, I stated there was no way I could or would pay such an extortionate sum – more than it had cost to ship the stuff from England to Brazil. I threatened that I would go straight to the British Embassy to report him for fraud. At first he tried crawling, rubbing his hands, trying to appeal to a side of me that didn't exist, then he became stroppy and abusive.

He knew I was powerless, though, because he had all the paperwork. But I didn't know just how powerless. I arranged to go with him to the Customs office. I honestly thought he was pocketing most of the money for himself. His car was outside, a taxi. 'He's no more a *despachante* than Ronnie Biggs,' I told Gordon after I had come back from the port. The Customs officer wouldn't even see me. His secretary told me in broken English that Senhor would only deal with my agent. She also informed me that the goods in their deposit were being charged a storage fee of $50 a day. So Portuguarra received his $2,500.

I put the episode down to experience and ploughed on with Gordon and Antonieta. I was determined this film project would come good and that we would enjoy it one way or another. Then Daniel bought the long-range radio for the boat, our only safety-line. It was the wrong type. I went to the shop to try and change it, or get my $1,600 back, but it wasn't that simple – nothing is in Manaus. The shop owner wouldn't refund the money, and he didn't sell the type we needed. I was even more frantic with worry over finances.

On 21 October, as planned, Chris returned to Italy, Daniel to Argentina and Eric to Denmark. It was five days before we were due to return to the giant terrapins on the Rio Branco and Chris told me we were in the capable hands of their partner, Plinio. I was filled with dread.

The sight of the scaffolding being fork-lifted from the container lifted my spirits briefly, but after loading the seven tons onto the *Brandão,* they came tumbling down again. Avoiding eye contact, Plinio told me that the boat needed further repairs – something to do with the engine. So yet another delay. I decided to sleep on board and try to keep an eye on things. The nights at the port were just as busy and noisy as the days. People hustled and bustled, loading and unloading boats. You needed eyes in the back of your head to make sure nothing was stolen. Boys no older than ten wandered around selling ice-lollies in the sweltering smelly heat; at night they slept under lorries on the dirty ground. Human excrement and rotting fruit floated all around the boat. Ghetto-blasters at full volume competed with each other from all sides.

We finally escaped from the port ten days later than planned, on 31 October at 10.20 pm. Gordon and I sat on the upper deck, opened a few beers and, laughing, bade good riddance to Manaus.

Three hours later, with the lights of the city still visible, Plinio steered the boat at full speed into a sandbar. We were stuck there for the night. The following morning, all our efforts to pull the massive hull free failed. By mid-morning we had summoned help from another boat, which succeeded in pulling us free. This time, Gordon and I decided not to tempt fate again by opening a celebratory can of beer.

We passed the following four days by painting every inch of our tons of scaffolding in green and brown, so that it would be camouflaged among the trees when it was erected. In the hot sun it dried quickly, which was just as well. The *Brandão*'s draft was simply too deep for the Rio Branco at this time of the year. We were constantly shifting the back-breaking metal poles from one end of the boat to the other in an attempt to shift the weight of the boat whenever she became stuck in shallow water – which happened often.

On 4 November we called into a small village of 30 or so timber huts built in a straight line along the high river bank. There were very few people about, and while Plinio's short, thick frame climbed up the ladder to the community, I watched several women washing clothes in the muddy river water, their children splashing about next to them. This was where Shagger lived. Shagger, we had been told in Manaus, was the person to escort us through the difficult shallow channels of the Rio Branco. He was an experienced *prático.* Watching the gangly, middle-aged man swaying his way down the ladder did not fill me with confidence. He was obviously drunk.

Plinio explained that it was a *festa* day, and he had been drinking. Could things get any worse, I asked myself, and returned to the paint pots.

With Shagger incapable, and still topping up with *cachaça* provided by Plinio, our boat had got stuck again and we were back in the water. My temper was rising by the second. At one stage, trying to shift 70 tons by human effort alone, the propeller sheered through the rudder chain. As I wondered when this nightmare would finish, lady luck smiled on us. Two barges – *balsas* – were making their way up the Rio Branco, and would be going past the Sororoca beaches. On board the leading *balsa* was an experienced *prático* who offered to tow us through. As an added bonus, the barges were loaded high with thousands of crates of beer, and after a few bottles we didn't even care about the *piums.*

We passed a small community called Santa Maria, the last sign of civilisation marked on the map. It was just a bundle of wooden shacks on the east bank of the river. Plinio emerged from the helm and told me there was a bar there where one could buy alcohol, but I said no. I felt alcohol had caused enough trouble and, moreover, we were only two hours from the nesting tree of the jabiru storks. I remembered speaking to Francisco, who lived 200 yards further along the river bank, and how, on our first meeting in February, he had told me proudly that a pair of storks had nested there for 17 years. He had seen them come and go every year. He talked as though he was their guardian, and he also told me that they would be nesting in November.

An hour beyond Santa Maria the two *balsas* came to a halt. The river was too shallow for the *Brandão* to continue further. There was nothing to do but go for it with our fast boats. We stopped off at Francisco's, who was sitting in front of his thatched hut with three young children at his feet. He was showing them how to make a fishing spear. He was sad. Five days ago a passing fisherman blew one of the birds out of the tree with a shotgun. The other had disappeared. Once again, I felt a sensation of powerlessness come over me.

It was four o'clock in the afternoon of 5 November and as I sat there, depressed, I thought about home; of families huddled together, watching Guy Fawkes night bonfires and fireworks, eating baked potatoes and sticky treacle toffee. I would have given anything to have been there, a million miles away from here.

At midnight, unable to sleep, I paddled across the river from Francisco's hut and sat on the soft white sand. The river was silvery calm under a moon that was almost directly over my head. The giant 150-foot high samauma tree, standing head and shoulders above its neighbours, was silhouetted perfectly against a pale cerulean sky, the huge outline of the ancient nest clearly visible on a thick bough. The birds should have been there, but were not and never would be again, because of some senseless human.

At 6.30 the following morning we bade farewell to Francisco and his family. We were to continue by fast boat to the turtle beaches, a frightening 120 miles away across the equator. As we pulled away from the river bank, with a sober Shagger on board to guide us, I couldn't help looking one last time up into the tree, and wondering what might have been.

The journey north took seven hours and this time I remembered to wear a hat. Shagger was useless, finding channels that even the shallow metal canoe couldn't pass. Worn out, we arrived at the IBAMA hut and there was Paulo, predictably still looking like the world was about to end. Despite his countenance, Paulo *did* have a passion for the terrapins. His job there was a largely thankless task, and sometimes dangerous. I liked him most of the time but I don't think, deep down, he liked us.

No sooner had we arrived than Gordon had to return south with Plinio, all 120 miles again, to make sure that the *Brandão* moved to deeper water. Otherwise, as Paulo told us, with the river falling the way it was this year, we might be stranded for six months. Gordon could also bring back the film equipment that we had been unable to fit in the boat on our first trip. The adult terrapins were not expected to come out to lay for another week or two, so there was enough time. I would wait for his return and, in the meantime, shoot the sequence of landscapes we needed to set the scene. They left just before dawn, Plinio with a face like thunder. I was onto his game. I had no doubts that he was up to something but I simply had to wait for the proof.

Gordon and Plinio arrived back on 11 November. I had some good news for Gordon. Exploring the area I had seen a lone jabiru flying low over the tree tops and swooping down into the spacious crown of a samauma tree. There to my utter surprise and excitement was another bird. A pair on the nest. This was special, because in northern Amazonia they are extremely rare birds. From the look of the nest it was a traditional site, with many layers of sticks piled up on top of each other year after year, like an eagle's eyrie.

Unfortunately, Gordon had nothing but bad news for me. When he and Plinio had arrived back at the *Brandão,* they found one of the two beer barges tied up to the river bank a few hundred yards from our boat. They stopped to say hello to the crew who had helped us through the shallow waters and were invited to return later that evening to play dominoes. After dinner Plinio motored off alone to meet them, Gordon preferring his hammock. Fifteen minutes later Plinio returned, shouting for help.

Gordon went with Plinio to the barge, together with Antonieta who had remained on the *Brandão* to keep an eye on things. As they climbed aboard they almost tripped over a prone figure. A man was lying on the deck with a large kitchen knife sticking out of his chest. The man who had killed him was one of the barge's crew and, after the stabbing, he had run off into the forest. According to the rest of the men on board he had been well drunk and was bragging that when he got to Boa Vista he was going to kill a bloke who had done him a bad turn. His fellow drunk scoffed and said to him that he wouldn't have the courage to kill a man. The conversation went along simple lines... 'Yes, I would.' 'No you wouldn't.' Drunk friend repairs to galley and returns with bread knife. 'Yes I bloody well would' ... 'Arggghhhh!'

Gordon and Antonieta decided to get back to the boat immediately and to move it away from this area. Ten minutes later, and before Benedito had time to start the engine, the sound of shouts and a scuffle were heard from the *Brandão*'s bow. The murderer, now clutching a machete, was trying to climb aboard our boat. Shagger grabbed the shotgun, which we always kept at the helm. The murderer, seeing the weapon, leapt off the boat and had reached the treeline when Shagger let loose. The man was blown off his feet and crawled out of sight into the trees.

Plinio steered the boat full ahead, and five minutes later hit a tree trunk lurking just below the surface. It sheared off two blades of the propeller and shattered two planks. Under the water line the river water gushed in, rapidly filling the 70-ton hull. Until three o'clock the following morning the whole crew, weak from fatigue, frightened by the prospect of a maniac on the loose, battled in the bilges of the *Brandão* to prevent her sinking. Well, almost the whole crew – Gordon, exhausted, slept through the entire episode.

Chapter Twelve

Charge of the Tartarugas

It was six weeks before Christmas when we first settled ourselves on the vast sandbank to wait for the giant terrapins (*tartarugas*) to emerge. Almost unbearable daytime temperatures of over 40°C made working extremely difficult. The searing heat of the equator was draining and sometimes it was an effort simply to walk. The *piums* plagued us and in the end we resorted to wrapping towels around our heads for protection, leaving slits for our eyes.

With more than 4,000 adult terrapins in the water around the sandbank I felt confident that we were going to get the images I wanted. We knew they were frustratingly timid, and would only come out to lay their eggs at night. So we took special precautions to prevent the terrapins from seeing us, and dug pits in the sand so that our heads were level with the surface of the sandbank. We peered over the edges of our pits like soldiers waiting for the enemy to attack. The stridulations of crickets and the distant cicadas in the forest edge added to the sensation that something was about to happen.

Every night we waited in our pit for the unmistakable sound of a giant terrapin, shell squeaking, lugging itself across the sand – but it never came. Almost three weeks passed. Paulo just thought I was impatient, and simply said we had to wait.

We wanted to return to Manaus for Christmas, and were under pressure from Gordon, whose girlfriend Sandra was coming from Glasgow to visit. 'Her mother will kill me if I'm not there to meet her at the airport,' he told me.

'They will be out laying any night now – don't worry,' I said, knowing that I couldn't leave the area until I had the film in the can, even if it meant Christmas in this lonely outpost. I also knew that it would make things much more difficult for me if Gordon were to leave before the beasts nested – and things were hard enough as they were.

The days and nights dragged by. We spent our days under a small plastic tarpaulin on the riverbank in the insufferable heat, attacked by blood-gorged *piums*, with decent sleep beyond our grasp.

'Perhaps I could hitch a lift with a passing fisherman?' Gordon asked one morning. With only two weeks to go until Christmas panic was well settled into his mind. As a contingency, I agreed it was a good idea, and we spent hours calculating how long it would take him to get to Manaus. With the *Brandão* we needed four days, but with a *batelão,* a small fishing or trading boat, it could take much longer. We also knew that passing boats were thin on the ground.

Then, one morning, Paulo returned from his dawn search of the more distant sandbanks with the news that on Acaituba he had discovered five *expansa* nests. It was just what I had been waiting to hear. Without delay I decided to shift our attempts to that sandbank, nearly four miles south of Sororoca. During the day, with Plinio and Francisco (an IBAMA worker), we transferred the equipment and dug new lookout positions on Acaituba. I left Gordon resting in the camp, telling him I would see him later in the afternoon.

On Acaituba I looked at the five nests and felt rising excitement. At the river-edge I followed the tracks that one female had made the previous night. The symmetrical patterns made by her feet wandered for 100 yards, at times crossing the track made by another. They were the only marks upon the sand and they looked as though an amphibious craft had left the water to explore the sandbank. The only indication that something else had happened was 250 feet into her journey. There the surface of the sand over 6 square feet had been flattened out after she had filled in the nest to cover her eggs. I knew that if I dug down three feet or so I would find over 100 ping-pong ball-shaped eggs. For some reason, after she had laid her eggs, she had not returned directly to the river, but had flippered herself further into the heart of the sandbank before perhaps realising she was heading in the wrong direction.

I was convinced it was starting. We had a generator to illuminate the area of sand where our chosen terrapin would nest, but the lights could only be put into position once she had started to dig, otherwise she might be put off laying. Because of the noise the generator would make, we ran 300 feet of electric cable across the sand away from the site to minimise the din. All 300 feet had to be buried in case a wandering terrapin snagged the cable. By three o'clock in the afternoon we were ready to return to the camp. Although exhausted from toiling under the sun, adrenaline kept me on edge – I was sure that this night the *tartarugas* would come.

'*Seu amigo sumiu!*' Paulo thrust his bottom lip out – compass direction south, and laughed. 'Your friend has disappeared.' It didn't sink in for half a minute. The language was still beyond my understanding when spoken quickly. He explained that earlier that day, just before noon, a small trading boat had pulled into the IBAMA post and Gordon had asked them where they were heading. Manaus, they told him, and he was gone. He left me a note saying how sorry he was, but he felt he had no option. The fear of not being there to meet his girlfriend had made the decision for him.

Loneliness descended upon me. The *piums* seemed worse and the approaching night darker. Trying to explain to Plinio how to erect filming lights, what to do when things started happening, seemed an impossible task. I had to mime almost everything. He watched, resenting my instructions. I knew he had no comprehension of how quiet we would have to be. As Plinio was gathering up some gear to head out to the sandbank I looked at the tattoo on his back. Clumsily home-made, the torso of a naked woman stared at me.

In the depth of our sandpit Plinio shifted and shuffled and then, like a camera flashlight being fired, lit a cigarette. I pointed angrily for him to extinguish it, which he did.

The night was silent. I strained to listen for the sound of a terrapin. Just after midnight, with a half-moon casting a pale silver-grey light over the sand, and with Plinio asleep, I heard it! Unmistakable, the heavy weight of a female terrapin using her powerful feet to pull herself across the sand. The underside of her shell scraped the floor – like a damp leather across glass – as she moved forwards in surges of six or nine feet. She would stop and listen – I listened too – then move forwards again. The sound was coming closer.

I nudged Plinio, put my finger to my lips in a shsh sign and thought of Gordon somewhere downriver. The two film lights were on their stands about ten feet away from us and the camera was ready on its tripod. Their metallic shapes looked weird and out of place in this wilderness.

The *expansa* terrapin moved again, closer to us, and then I saw her. She had stopped about 15 feet away and I watched as she lowered her huge head and seemed to sniff the sand. We both heard the next one coming, and my heart began to beat harder with expectation. The second terrapin almost bumped into the first, but carried on past the spot where we were dug in, and out of sight. Within the next 15 minutes she would start to dig. Then four more *expansas* came towards the same area in front of us. They looked like a prehistoric raiding party, the advance landing group of an army coming to take over the sandbank. I smiled to myself, watched and listened. How remarkable it was to be here, a witness to it all.

The terrapin's front feet scooped at the sand, throwing it behind her with powerful thrusts. I had made my mind up to wait until she had dug herself below the surface before attempting to film her. The *expansa* excavates a deep nest, usually about three feet down. Their choice of site depends on several factors, including proximity to the water, but also being high enough above it to avoid flooding should the river level rise unexpectedly.

It was time for us to move as quietly as possible. I motioned to Plinio to go over the back of the sandpit, walk over to the generator, and switch it on after five minutes. I crept up and over the front, trying not to make a sound. Then I turned the camera slowly to point at the place where she was, now head down unable to see me. The scooping of sand had stopped – she had heard me. Suddenly the area was bathed in light –1200 watts of it. The generator was a distant humming sound. I zoomed in with the lens and focused on the hole, then zoomed out back to a wide angle. Providing I didn't move the camera, I could now choose any focal length on this big lens without re-focusing.

I waited and prayed that she would begin digging again. In the wide pool of light I immediately saw three others that had arrived and dug nests, somehow without me hearing them. One of them had her head poking out of the hole she had dug and was now obviously laying her 100 or so eggs. Her head rocked slowly back and forth as she watched me. She was too far away from me to film. I froze for ten long minutes before the female closest to me began to chuck out lumps of sand again. I breathed a sigh of relief.

It took almost two hours from start to finish. After laying all her eggs, she moved upwards and began to scrape the sand back into the large nest hole. When it was almost filled she rotated her massive frame to pack the sand down over the eggs. What a struggle it will be for the little ones to get out of that, I mused, as she set off back to the river 50 yards away, squeaking and stopping even on the return journey. I moved the camera closer to another nesting terrapin, and managed to get close-ups of her head. As she laid her eggs and, like the previous one, wobbled her head, I noticed dark stains under her eyes. She seemed to be crying and I wondered if the relief from discharging more than 100 eggs caused this.

The following night I was back on the sandbank, but this time in a new pit closer to the edge of the river and with a different assistant, Plinio having made his lack of enthusiasm clear. Francisco was about 25, wore ridiculously baggy camouflage trousers, and seemed a likeable chap. Before we had left the camp I explained as best I could what I wanted him to do – keep very quiet, don't make footprints in the filming area, signal for switching generator on, etc, and he had nodded and smiled at me.

That night the terrapins came in their droves across the four sandbanks that make up the major part of the *expansa* nesting beaches. They appeared in waves of ten to 15, and each would take up to two hours to choose a nest site, lay eggs, refill the hole and then return to the river. It is one of nature's great spectacles. Our new pit was perfect and Francisco and I both watched in wonder as groups emerged around us, jostling for a space to dig. One huge female almost slipped into our hollow as she tried to avoid two other giants already starting to excavate only nine feet from us. Before four o'clock that morning 200 females had nested, and I had the sequence in the can. It was a fantastic feeling.

'*Nascer de sol*,' Francisco whispered to me, his rusty rifle clutched to his chest. The sunrise – or birth of the sun in Portuguese – came at ten minutes to six. The smooth sands of the previous day were gone, replaced by just craters and humps, as though a battle had taken place. Nature's undertakers, the vultures, swooped down from the trees at the forest edge.

Some *expansas*, digging too close to an existing nest, had scooped out the eggs below. These accidentally uncovered eggs littered the sand and attracted the carrion eaters. In flight vultures are magnificent, soaring gracefully on thermals with primary feathers extended like an eagle. But on the ground, or sand, they are ungainly disagreeable creatures that seem to hobble and squabble. Their sharp beaks snapped and gorged greedily on the fresh eggs. Protein-rich liquid spouted into the air among clouds of *piums*.

After I had filmed this hellish feast, Francisco said he was going to look for a duck for us to eat later. He trudged off towards the forest 200 yards away to the east. I was daydreaming; watching Francisco quarter the forest edge where the ground dipped, creating hollows and pools. Suddenly the few remaining vultures took flight, squawking noisily. It wasn't Francisco who had disturbed them. There was a sudden stillness – not even the crickets stirred. After he had wandered some 50 yards or so, head down, rifle in hand, something caught my eye in the edge of the trees behind him. The hairs on the back of my neck tingled. I could hardly believe what I was seeing.

A jaguar jumped down the rise where the forest fell away to the sandbank and, with her head hung low, sniffed the tracks that Francisco had just made. It was my first clear view of a wild jaguar – and it was following Francisco. I don't know why I didn't feel any fear for him, perhaps I was simply overawed by the sight. *Panthera onça*, the spotted jaguar, looked magnificent in the early warm light of the day. She was lean and powerful, her markings dazzling, the black rosettes highlighted on her golden coat – and the one animal all people who live in the forest fear above all others. In this business of making wildlife films 'sod's law' is omnipresent. I had used the last can of film on the vulture's feast and I could only wince inwardly at this lost opportunity.

Francisco turned west away from the forest and started to walk towards a small forested island 100 yards to my south. The hunter being hunted, I thought briefly. The jaguar didn't alter course, but followed Francisco's tracks along the edges of the pools. The usually noisy vultures were still silent. I could have shouted to him but I knew what would have happened – the rifle would have been raised and fired. I honestly didn't believe that the cat would attack him, and if I had seen it start to charge I would have screamed to warn him. The jaguar was only 50 or 60 yards behind Francisco when she turned away from the pools at exactly the point where he had altered course. She continued to shadow his tracks. He eventually turned to face me, and waved. He never once looked behind him. Coming towards me now, with the cat keeping a steady 40 to 50 yards behind Francisco, I decided I would wait until he reached our pit before pointing at the cat.

As I watched this magnificent cat's easy stride I thought back to my two years in Guyana filming *Giant Otter* for Survival. My producer at that time, Caroline Brett, and I had talked eagerly about making a jaguar film, their elusive nature making them the most difficult of all the big cats to film. Making that film has been a dream of mine ever since. Nine years later, on this Rio Branco sandbank, with a wild jaguar coming towards me I could see the dream turning into a reality.

Francisco plodded over the edge, sliding down the sand into our pit. Grinning from ear-to-ear I put my hand on his shoulder, pointing to the jaguar now only 50 yards away from us. He began to smile and then he saw the jaguar. In a fraction of a second the smile was replaced with a look of wide-eyed fear as he wrenched the rifle upwards towards the cat. Just as quickly I pulled it down. It took a few seconds for him to register what was happening. We were both looking at the jaguar when it stopped. It stared straight at us for a few seconds, so remarkably close, and then it simply turned away and padded off towards the forested island. I said '*fantastico*' to Francisco, slapping him on the back, and he seemed to understand. It was a good guess – perfect Portuguese!

The first part of our filming accomplished, we made ready to set off for Manaus – hopefully in time for Christmas. Saying goodbye to Paulo and the *piums*, albeit temporarily, was a good feeling. Sadly Francisco had come down with malaria fever the day after our success on the sandbank. He looked in a bad way and I left Paulo with a dose of Fansidar, malarial medicine, to give to him. Plinio and I departed armed with a tatty piece of paper from Paulo filled with demands for 'goodies' that he expected us to return with at the end of January: music cassettes, belt, pocket knife, binoculars, and a portable TV were prominent among the food items.

We stopped only once on that long return journey to the *Brandão.* Paulo had asked us to drop a bundle of dried meat off at Acaí, a small IBAMA camp six miles south of Sororoca. Two young men kept vigil there and welcomed us eagerly. They boiled the familiar thick sweet coffee sludge for us and yapped away with Plinio while I, pointedly, looked at my watch. To be polite I also accepted a plate of food – fish and farinha – the leftovers from their last dinner. Within 20 minutes I had the most painful stomach cramps and asked where the latrine was. The pit was behind the make-shift camp and in full view of the men under the tarpaulin shelter. It was filthy and it stank. I told Plinio we had to leave.

The *Brandão* was tied up to the floating IBAMA post near the mouth of the Rio Branco. Sitting on the weathered floorboards of the post, in a rickety old rocking-chair, was Chico Preto (Black Chico). He had a small dirty white mongrel dog at his feet, a smile on his face and a thick home-rolled cigarette in his mouth. He looked totally at peace with the world and I envied him. He chatted to Plinio while I climbed aboard the *Brandão* and organised my hammock.

When Plinio came aboard he told me that half an hour after we left the two men at Acaí one of them had gone to the latrine. While squatting there he had been bitten by a snake, a *fer de lance*! His friend had paddled him to Sororoca and Paulo set off in their fast-boat to take the young man to Boa Vista by canoe, where he could be hospitalised. Boa Vista was five hours away from Sororoca and I wondered if they would make it. I felt a huge sense of relief when I considered how close the snake might have been lurking to my own buttocks earlier that day.

Homesickness is something I have learnt to control over the years of making wildlife films. I have had to. The distance from home, the loneliness, is never far from my mind at the best of times but it was in 1985, in China, that I developed a mechanism to deal with it. There, in China, before the country had begun to open its doors to foreigners, I had never in my life felt so desolate and homesick. I was isolated from everything I had been used to: no family or friends, no method of communicating with the people around me, no telephones, and a culture that couldn't have been more alien. One dreadful night, after witnessing an execution in a rural community a million miles from comfort, my feelings overwhelmed me and I lay on a mattress in my cell-like room and sobbed.

I was only one month into my first wildlife commission, a five-month contract. The following morning I gave myself a bracing lecture. I told myself that failure was out of the question and that I couldn't go home even though I wanted to, so I had better get on with it and be grateful that I was doing what I loved – filming wildlife. My one-man pep-talk did the trick. Although there have been times since when those old feelings have started to surface, I always think back to China. If I could survive that, I could survive anything.

Remembering my time in China, Manaus didn't seem half as bad. Gordon – yes, we were back on speaking terms! – and Sandra were there, and so was Antonieta. We scoured the few decent shops for presents and wrapping paper and laughed at the sound of piped Christmas music

Brazilian style – *Silent Night* to a samba beat just isn't the same! We had decorated our apartment walls with the dozens of cards we had received, all depicting cold snowy landscapes that we daren't tell each other we hankered for. Antonieta had never even seen genuine snow. A tasteless plastic tree was the best we could find, then, just in time despite the Brazilian postal strike, our 'Red Cross' parcels arrived from home. Emma had sent a Fortnum & Mason's Christmas pud and Harrod's Christmas crackers, which survived the Brazilian postal system, together with home-made mince-pies, which didn't. Tina Schofield and her family in Norwich sent a parcel that contained a complete miniature Christmas, including little malt whiskies, that delighted us no end.

On 25 December, 1992, we opened each other's presents. Gordon and I approached each other with suspiciously similar looking parcels. Small oblong boxes, five inches long, even wrapped in the same patterned paper. We had both bought each other the very same model of Swiss Army knife. 'Great minds think alike,' Sandra commented.

After the turkey – a frozen import – I paraded into the lounge, side-stepped the chunks of plaster that had fallen from the ceiling, and plonked a flaming Christmas pudding onto the dining table. To the resounding cheers of my companions we watched the blue light flicker and die away. It was the beginning of the rainy season, the humidity was 100 per cent, the temperature 35°C.

Christmas in Brazil is a national holiday. I suggested that Gordon arrange some sight-seeing trips for Sandra, jungle tours being out of the question given that we both knew what swindles most of them were. Among the many pamphlets he had gathered was one for a place called Eco-Park. It was an hour by boat and, according to the brochure, it featured free-ranging but tame Amazonian animals that had been rescued from poachers.

The Eco-Park project was run by a Dutch couple, Marc and Betty van Roosmalen, as a collaboration between themselves and IBAMA whereby they took in many orphaned creatures that had been confiscated by the authorities in Manaus. I had already heard of Marc through his work at INPA, the National Amazonian Research Institute, where he was a senior scientist in the botanical department. Marc and Betty had lived in Manaus for seven years and before that Marc had been studying spider monkeys in

Surinam, north-east South America. Gordon suggested it would be worth meeting Marc as a possible source of tame animals for filming.

In my experience most people who live long-term in these far-flung corners of the world are a little 'off the wall'. I include myself in this generalisation, and Marc and Betty certainly fitted that picture. They were both in their mid-40s and looked like 1960s hippies. Marc was tall and spindly with shoulder-length blond hair and pointed features and seemed shy. Betty was more outgoing and laughed easily. They were both extremely friendly.

Their house was in a guarded complex called 'Garden of Europe', on a road called 'Country of Wales'. It occupied two lots, and had a large back garden. The grounds overflowed with tropical palms and plants, with orchids sprouting from tree forks. A noisy colony of caciques nested overhead, the bright flashes of yellow feathers among black darting to and fro as the birds collected strips of palm to weave their homes. A baby red brocket deer grazed on the coarse grass with two of the largest tortoises I have ever seen. A pair of scarlet macaws and many parrots squawked noisily, one parrot mimicking the sound of a telephone ringing. A splash in the swimming pool drew my attention and a southern river otter gambolled in the clear blue water. I half expected to see an ark being built in some corner.

We talked late into the night about our respective work and experiences and, inevitably, I mentioned the difficult time I was having with the Rio Branco project. Marc and Betty both laughed knowingly. At that time, unknown to us, they too were struggling to keep things going with their project. At the end of a pleasant evening we promised to keep in touch. 'Who knows? Perhaps we could work together in the future?' were Marc's parting words.

Hogmanay turned out to be a lively celebration. We had met an army vet, Gino, at the military zoo in Manaus. He spoke reasonable English and had a great sense of humour. Thirty-one years old, with a military crew-cut and an insane grin, we were to become close friends. He invited us to pass New Year's Eve with him and some colleagues at his house in the suburbs. I finally took the plunge and asked Antonieta out. An odd feeling – something I had not had to think about for 18 years. The combination of a barbecue, Gino on guitar, fireworks and too much drink made the evening swing. At eight o'clock – midnight in the UK – the telephone rang and Gino called out, 'Some crazy guy's on the phone for you.' As I put my ear to the receiver I could hear *Auld Lang Syne* being chorused. I knew exactly where it was coming from. The singing died down and a great cheer went up, 'Happy New Year, Nick!'

'Hello Partner – how are you?' It was my good friend and business partner from Lancashire, Nick Peake. He and his partner, Mary Rose, had gone for a few days' break to the island of Mull and were staying in my house. 'Before you throw a wobbler, we're not having a party in your house, we are over the road in your neighbour's, Rob Barlow.' It was a wonderful and unexpected surprise, which cheered me up even more than Gino's Brazilian beer.

The holiday had passed all too quickly. Sandra and Gordon had a tearful parting at the airport, and we were driven by the need to film the second stage of the terrapin sequence. Somehow, against all odds, we mobilised the *Brandão* and assorted crew for our departure to the Rio Branco on 11 January, 1993. Five days later we were among the *piums* of Sororoca once more.

Paulo met us in Sororoca with his hand outstretched – but not to shake mine. His 'wants list' was received without a word of gratitude and I thanked my common sense for not having brought the binoculars he had requested. Francisco, happy to see our return, had recovered from malaria, but the man bitten by the *fer de lance* snake was still in hospital, gravely ill.

The river's water level was rising rapidly and there was a possibility that some of the terrapin beaches would be flooded. This does happen occasionally, wiping out thousands of nests. At dinner that first night, in between shovelling spoonful after spoonful of feijão beans into his mouth, Paulo told me he would show me the records he keeps every year for the hatching times of individually marked nests. The first were due to hatch within our first week. As there had been a staggering 4,000 nests excavated, I didn't feel we would have to wait long. As soon as the hatching was on film, we needed to build a scaffold tower in preparation for the next season's jabiru nesting, and then, finally, to make our way to the Xixuau to begin shooting the main section of the film, the wildlife of the creek.

Rather than have the *Brandão* moored at the nesting beach, we decided to move it a little further downstream and make a 20-minute journey each day by fast canoe. That way we could keep away the rabble on the boat. The first morning we filmed back on the sandbank of Acaituba. I showed Gordon where the jaguar had appeared and followed Francisco. The air was thick with the mixed scents of flowering trees, the brilliant blue sky spotted with puffs of cotton wool cloud. The sandbank, now pock-marked with hundreds of nest sites, was still a haunt of the vultures. A few of the nests had caved in, forming a cone shape, where the baby terrapins had already hatched and left.

It was frustrating at first, because the only sign that a nest was hatching was when a vulture or other bird of prey would swoop down and grab a fleeing terrapin. This left us no time to waste. We chose an area where seven nests were closely placed to each other and simply waited. On the third morning of our vigil the hatchlings emerged.

It was 5.40 am, 20 minutes before the sun came up, but with enough daylight bathing the landscape for me to film. Gordon shouted 'here they come' as more than a dozen miniature *expansas* ran for their lives straight towards the river. They possess an amazing sense of direction, considering they cannot possibly see the water from the dunes where they emerge. Their tiny flippers clambered furiously over the sand, propelling them forward at an incredible speed. Black-faced caracaras and vultures began to dive-bomb. Relatively few terrapins were taken, the majority reaching the river safely.

Where the sandbank met the river an extremely steep slope cut away our view. On our last day of filming the hatchlings, we set up the camera at river level to catch them coming 'over the top'. This was to be one of my favourite sequences in the film *Creatures of the Magic Water*. Hundreds of hatchlings appeared on the brim of the hill and scrambled down it for all they were worth, finally slipping into the river. It was astonishing to think that three years later some of those tiny creatures would themselves return to that very beach to lay their own eggs. Nature *is* astounding.

Thrilled by the morning's display on the *tabuleiro* (sandbank), Gordon and I planned our following day's work as we motored back to the *Brandão*. Once again, our spirits were to be dashed within minutes. I climbed aboard the bow and crawled through the diesel store where one of the crew slept on a makeshift cushion mat. To my horror I saw, nestled in a crumpled bed-sheet on the floor, about 50 pale white round eggs. Here we were, in a protected area, representing Survival and the Amazon Association, and some idiot had robbed an *expansa* nest while we were out filming.

I shouted for Plinio in a furious rage. There was nothing he could say – he knew exactly who had done it and had not had the strength to object. I sent Gordon to re-bury them, knowing they were probably already dead. I was determined to ditch those responsible as fast as it was practicably possible.

The mood on the boat was disagreeable after my flash of temper. Plinio kept well out of my way except when he needed to know about travel arrangements. Gordon and I immersed ourselves in the tower building so that we could move on to the Xixuau. Finding a second jabiru nest was a stroke of luck, but it meant two days of hard labour lay ahead. A 150-foot tower weighs about three tons. We had to transfer our scaffolding by hand into the canoe and paddle it to the forest edge. The metal poles then had to be carried 500 yards to the base of the giant samauma tree and, after careful levelling of the rough ground, erected.

All hands helped to shift the aluminium poles to the site, which was unexpectedly guarded by a big snake. A glossy black and red head reared and hissed angrily at us. We hadn't seen this type before and couldn't tell if it was venomous. We tried to shift it with a cut pole, but understandably it wasn't a happy serpent. Eventually with most of its six-foot length wrapped around the stick, we shifted it 20 yards or so away. Within minutes the persistent creature was back. 'Whoa,' I heard Gordon shout as he almost walked over it. This time we moved it much further into the forest and there it remained.

During our African expedition, Gordon and I had built more than 20 scaffold towers for filming. We used to joke that if one day we couldn't make wildlife films any more, we would be welcomed with open arms by the supply company in London. The Department of Health and Safety might have had something to say about our methods, of course, teetering aloft without safety lines or harnesses. With all the necessary metal to hand we could erect 90 to 120 feet in a day. It was back-breaking work, but satisfying. As you gain height, each stage offers a different view of the forest's layered structure, through lower to middle storey to canopy, and in some instances above it. Flowers and fruits, invisible from the floor, suddenly came into view. Birds approach to take a close look, and where metal touches tree – ants arrive.

The crown of this magnificent samauma tree opened out at 70 feet and then spread its enormous boughs skywards to about 150 feet. It was exhilarating to be able to see above the surrounding rainforest. Stretching away north and south from my lofty position was the Rio Branco; the terrapin grounds, about two miles away, were clearly visible. We positioned our top deck level with the sprawling jabiru's stick nest at 75 feet. I fixed the angle bracket and pulley on the topmost pole and shouted down to Gordon to hoist up the film hide. As I unwrapped the bundle of olive green material and spread the poles out I thought how extraordinary it would be when we were finally installed in it, looking out on to the nesting storks.

As I tied the final guy rope to a metal pole there was a loud whooshing noise. I froze. Less than 25 feet from me, elegant and ugly, was a jabiru stork. I crept into the hide and peered through a screened flap. It didn't seem bothered by the contraption we had placed there, and so I looked over the back of the tower, where the bird couldn't see me, and waved madly at Gordon. He had his thumbs up – he had seen it, too.

Good assistants are worth their weight in gold, someone should have once said. Gordon knew exactly what to do, and 20 minutes later, having told the boat to move over to the other side of the river to minimise the disturbance close to the tree, he hoisted the camera box and tripod up as silently as possible. It was impossible for him to climb up because the stork would have seen him and taken flight. An hour later I had the start of a sequence in the can. As the sun dipped lower in the sky, and I wondered how long I would have to remain hidden, another whoosh announced the arrival of a second bird. They were obviously going to roost here for the night, and I would have to wait until after dark before lowering the gear and leaving the tower.

'Perhaps things are looking up,' I said to Gordon as we tried to find our way out of the tangled undergrowth by torchlight.

'Perhaps peccaries will fly,' the canny Scot replied.

Mid-morning on 20 February, 1993, we arrived at the mouth of the Jauaperi river. I was in my hammock leafing through the piles of research papers on manatees, terrapins and other aquatic mammals while Gordon was deeply engrossed in George Orwell's *1984*. The *Brandão* came to a halt mid-river. Without warning the fast boat roared away from our starboard side. Plinio was at the wheel. 'Where the hell's he going,' I shouted. '*Que isso?*' I asked Benedito. With sweat pouring down his face, oily rags in hand, he smiled, shrugging his shoulders. My blood pressure rose and remained at boiling point for two hours, until the sight of Plinio returning took me over the edge – almost literally.

I leaned over the side and grabbed the painter from the fast boat but the idiot hadn't taken it out of gear – and almost pulled me into the river before he realised his mistake. In a furious voice I shouted '*Onde você foi?*' (Where have you been!?) The man was drunk and aggressive, and I wanted to hit him. I am not a violent person and normally avoid confrontations, but this was too much. Gordon intervened, 'Leave it, Nick,' and Benedito explained that Plinio had just gone to say hello to Valdemar.

'Obviously saying "hello" before noon means getting pissed on *cachaça*,' I muttered to Gordon. I went after Plinio, to try and get through to him the seriousness of this project – but he was already comatose on his bunk.

Chapter Thirteen

Magic Waters

We had stocked up with provisions in Manaus in order to remain away from civilisation for two months, until the end of March 1993. Most of the food items we brought with us were non-perishable – pasta, dried beans and the like – and for fresh protein we intended to take advantage of the tasty fish that lived in the rivers and creeks around the Xixuau. I wanted us to settle into a routine while we were there and start to structure some of the main sequences of the film.

Our theme in the Xixuau was to be Carlitos's life and how he was teaching his young son Francisco to fish and hunt. Seasonal factors helped us to kick off, when Carlitos told us the toucans would be nesting soon. The site was conveniently close to Carlitos's house and we wasted no time in lugging more metal scaffolding to the tree. The nest, used year after year, was about 50 feet off the ground on the opposite side of the lake. After hacking our way through the undergrowth to make a narrow trail, we were covered in familiar blood-sucking tics. But at least we had completed our first day in the new location. That evening, armed with tweezers and torches, we spent almost an hour picking the tics off one by one.

Preparing the toucan's nest for filming was a delicate process. First we had to drill a series of holes and cut out a removable 'window' from the tree about 10 inches square. Unfortunately the tree was the home of several bees' nests – which we did not discover until I started to hammer a chisel into the bark. Within seconds we were cloaked in furious 'sweat bees'. It must have been the fastest both of us had ever climbed down a tower, and at the bottom we brushed away the hundreds of insects still clinging to our clothes. The only thing for it was what I called my 'no see ums'! I had scoffed at myself for buying them when, in Canada several years earlier, my sister Gill, an inveterate shopper, pointed out these anti-insect head nets. 'They'd be useful for you, Nick.' We may have looked stupid – judging by the crew's mirth we surely did – but they saved the day and we were able to continue working. Gordon, by my side, helped me to ease the panel I had cut out of the tree. We had noticed an unpleasant odour earlier, but the bees had taken our mind off it. The smell was there again and getting stronger. I told Gordon that I had read that toucans' nests were always fetid holes, but this was beyond fetid. As the plug of wood came away a stinking mound of black sludge fell onto our knees and legs, and then a mass of big brown flapping wings hit me in the face. Bats.

The stench was overpowering, but at least the bats had cleared up the mystery of the smell; it was their droppings. As I put the finishing touches to the hole, I vaguely remembered that I had read somewhere that breathing in the fumes from some bats' droppings could have a fatal effect. In went the final screw and I was down the tower in a flash.

Over the following weeks the boat crew idled while Gordon and I explored further from our floating base in search of giant otters. There were active breeding dens in the area, hides needed to be installed and the animals' movements noted. The filming of the otters was the easiest part for me, having followed them for almost two years in Guyana. All the above water shots followed quickly once we knew the area in which they were fishing. The hard part was going to be the underwater sequence. Fernando at INPA had a 'pet' giant otter and we had talked to him about the possibility of him coming with us on a trip to the Xixuau so that we could film her underwater. He agreed. He liked the idea and in any case wanted to place a researcher in the area to study the resident groups.

Using 'tame' animals in wildlife films for occasional sequences is more common than people realise. Many classic wildlife films wouldn't have been possible to make without their participation, and I wouldn't have been happy with the thought of being underwater with several six-foot-long wild giant otters – their teeth crunch through bones and skin with ease. However, filming otters underwater was still six months away.

In the event, we had to bring our return to Manaus forward because a fault had developed with our diving compressor. It was to be a quick visit – only five days. At the end of February we arrived elated by the film sequences we had shot but a little weary and ready for a few days' 'shore leave'. I had to get equipment repaired, send film back to Survival for processing, and resolve the situation with Plinio. But first I wanted to hit the only shopping centre in Manaus, appropriately called 'Shopping', eat a pizza, see a film and drink a few ice-cold beers. After soaking myself under a hot shower and smothering myself with talc and clean-smelling ointments, Antonieta, Gordon and I went out to paint Manaus red.

Antonieta wanted to continue with her college studies and so it seemed ideal to employ her as our 'fixer' in Manaus. Anti, as we came to call her, was well in control. She could hold the fort while Gordon and I were in the field.

Plinio was next on my 'things to do' list. Apart from his increasingly erratic behaviour, he had treated Antonieta like dirt from day one, and relations between us all were very tense. Plinio pre-empted things when he appeared at our apartment with '*mal notícias*' (bad news). The *Brandão* had mysteriously developed another serious engine problem and needed repairs costing $1,500. He had a receipt in his hand before the repairs had even been done! I didn't just smell a rat, I could see it. I told him to get the repairs done and come back. To cut a long story short, it was clear we could not continue the present arrangement and I immediately set about finding another boat that we could rent for the final eight months of the project. We spotted the *Natureza* tied to the river bank at Educanos in Manaus. She was floating – a good start – in a river of excrement and rubbish. Unlike the *Brandão* she was small, shallow-draughted, required less fuel and looked perfect for our needs.

'Pity we can't turn the clock back,' Gordon said as we looked over her. We met the owner and agreed a rental. Then we were introduced to her crew: Manuel, the young captain and mechanic, and Cabeça the mate. I explained to them what our project was about and Cabeça – 'head' in Portuguese – asked if we were German. Gordon laughed, and, as always, pointed out that I was a Sassenach Englishman and he was a Scot.

Seven days after our return to Manaus we were, once again, heading north to the Jauaperi river. This time it felt like an enormous weight had been lifted from us. The *Natureza* was faultless, the crew were hard-working and we enjoyed that first voyage on her like none other. Our arrival at the Xixuau was a cheerful event and the filming followed happily. In this eight-week trip we had much to accomplish. It was important for us to 'set the scene' and film a myriad of small things that would help the viewer to 'travel' to the location as they watched the film. The water, the leaves reflected in it, birds and butterflies – they all help to transport an audience to the place, and if a film-maker succeeds in that, he or she has indeed made a worthy film.

Carlitos came over early in the morning to accompany us on our travels. He would always be up at the crack of dawn to have a coffee and chat to the crew. Francisco, his 11-year-old son, was usually by his side. One such morning he asked us if we had heard a commotion in the small hours. He told us how the two black caiman who lived in the lake would, at night, swim in close to the muddy bank in front of his house and stare at the dogs. His mutts, spotting the predators, would immediately yap their heads off and inch closer to the water line. This night the unlucky one of them had got too close, and the massive caiman had lunged at it and dragged it into the water.

We had seen the two monsters from time to time, gliding menacingly across the narrower channel at the entrance to the lake. When Gordon and I went to swim and wash that morning, as we always did off the back of the boat, one of them – only its huge head visible – was mid-river about 300 feet away. We joked about it as we dared each other to dive in first. Gordon went, and I followed. We shampooed our hair and climbed out. Imperceptibly the caiman had moved to within 150 feet of the boat's stern – capable of getting to us within a matter of seconds. We both knew we would have to be careful in the future. They were *only* about 14 feet long, but they can grow to 21 feet. Their prehistoric heads looked fascinatingly gruesome as they swam from one side of the creek to the other.

Carlitos's principal daily work was to fish. He had a large family and fish was their staple diet, together with farinha. Farinha grain is made from the cassava root, a tuber, and the process of producing an edible product is hard monotonous work – something that Carlitos tried to avoid wherever possible. Sitting in a canoe, paddling slowly and pulling a few fish in was more his style and I couldn't blame him. It was while following him for many days in this way that we were rewarded with some truly special wildlife sights and sequences on film.

One afternoon we were faithfully tailing Carlitos while he cast his tarrafa net from the side when we spotted an anhinger perched on a branch low over the water. It is a diving bird, somewhat like a cormorant. Its plumage, however, is very different. Streaked with silver-grey and black, the anhinger has a fan-shaped tail, a long snake-like neck, beady eyes and a head and beak like a lance. It is usually found higher in the trees where it watches the water intently. It can spot fish with ease and its aim is deadly. It uses its beak to spear the fish underwater and then, on the surface, it expertly manipulates them into its gaping throat. When you watch one dive into the water it is easy to see why it has also been called a 'darter'.

This bird, only 150 feet away from Carlitos, was peering into the water – literally. Its head and six inches of neck were underwater as it looked about for fish. The black water of the creek was mirror calm and the sun, breaking through the gaps between the overhead trees, was back-lighting the bird, giving it a magical silver outline. It flopped into the water, disappearing from our view. We could see a trace of tiny silver bubbles coming to the surface showing us its path underwater. It returned to exactly the same branch and clambered out in an ungainly fashion. With a flurry it spread and shook its wings to get rid of the excess water. The droplets, like tiny jewels, cascaded through the air, fell and briefly caused the perfect reflection of the bird in the black water to break up. We stayed there more than an hour, up to our shoulders in the water, capturing the sight on film. It is one of the opening shots of *Creatures of the Magic Water.*

On another memorable day, in early March, we came across a sloth swimming the width of the creek. Of all the animals in Amazonia the sloth has to be the most peculiar. Sloths have long limbs, short bodies and a stumpy tail and the whole creature is covered in a pile of shaggy, grey, coarse fur. Their faces seem to wear a permanent expression of sadness. They generally live an upside-down existence, hanging from trees secured by two or three (depending on the species) extremely strong claws. These claws look like hooks and could inflict serious wounds on the unwary. In Portuguese they are called *preguiça,* which means lazy. It is an unjust title. Considering their odd appearance and cautious, rather than slow, nature they are surprisingly good swimmers. Today, like in Bates' day 150 years ago, few people are aware of their competence in the water.

> The inhabitants of the Amazonas region, however, both Indians and descendants of the Portuguese alike consider the sloth a byword for laziness. It is very common for one native to call another, in reproaching him for idleness, '*bicho do Embauba*' (beast of the Cecropia tree), the leaves of the Cecropia being the food of the sloth. ...In one of our voyages, Mr. Wallace and I saw a sloth (*B. Infuscatus*) swimming across a river, at a place where it was probably 300 yards broad. I believe it is not generally known that this animal takes to the water. Our men caught the beast, cooked, and ate him.

Our sloth was a female and there was no chance of her being eaten. She had a small baby on her back which was straining to keep its tiny head clear of the water, while the mother paddled her arms and legs, quite gracefully I thought, to get to the trees on the other side of the creek. As she reached the nearest tree, which was standing in several feet of water, she clamped her claws around the trunk and, as though in slow motion, rose out of the water. Her shaggy wet coat had a green hue to it, caused by the algae which grow in the coarse hair and help to camouflage the creature when it is in the tree tops. Another odd detail about the sloth's lifestyle is that it only descends to the ground, other than to cross a river, once a week. At this time it comes down to the base of a tree, digs a hole with its stiff short tail, defecates and then covers the hole up again before climbing back up to the canopy. Gordon and I stayed in our canoe watching her for an hour or more until she eventually disappeared into dense foliage.

Early one sunny morning during the second week of March we paddled with Carlitos and Francisco to film them fishing with hook and line. They make their own lures from bits of wood. While we were filming Francisco throwing the line, his lure snagged on a branch underwater. Without hesitating he dived in and released it. It struck me how fearless he was for his age and what a competent swimmer he was. I told them I wanted to film the part underwater when he released the lure and they laughed. They laughed even more when I put the diving gear on and took the lure back down to fasten it to the branch again.

Francisco was happy to dive in again and repeat the action, but the first time he did it my camera almost touched the lure, ruining the shot. When his shape appeared in front of the lens he pulled the lure free, looked into the lens and smiled! To their growing amusement we had to do it again. Before I climbed back into the boat I said to Gordon it would be worthwhile getting a shot of the underside of Carlitos's canoe as he paddled overhead. It was quite a technically difficult operation. We arranged a plan whereby Gordon would put his face into the water (with his mask on) and as soon as he saw my signal would wave to Carlitos to start paddling. Meanwhile, I was kneeling on the white sandy bed of the creek in just nine feet of water with the large camera housing pointing up towards the surface. I had to correct my balance because there was a slight current. Behind me the sand dipped into deeper water and darkness. I looked into that darkness for a few seconds and then tried to put it out of my mind.

I sized up several angles, focused, set the exposure and was just about to shoot when I felt a tickle on my left knee. I squirmed a little but didn't look down, assuming it was a leaf or twig being carried by the flow. I glanced up at Gordon and put my eye back to the viewfinder. The tickle came again but this time I looked down and, to my utter horror, between my legs was the large head of an anaconda – its forked tongue flicking out tasting my knee!

Time seemed to go into suspend mode. I turned to look behind me and could see the serpent's body trailing away down the sandy slope. It was about nine feet long. I should have kept calm, turned around and tried to film it, but I panicked, resembling a Polaris missile as I reached for the surface.

'Grab the camera, for God's sake,' I gasped at Gordon, spitting out my mouthpiece.

'What the hell's wrong?' he asked, genuine concern showing on his face.

'A bloody big anaconda is what's wrong and it gave me a hell of a shock when it put its head between my legs.' Genuine concern was now changed to spontaneous guffawing.

'It was right there between my legs!' I repeated as I put my face into the water to look for it. That was a mistake. The creature had followed me up to the surface and was shockingly close.

'Pull me in,' I yelled and with one super-human thrust from my fins and Gordon pulling on my diving tank I launched myself up and over the gunwale, collapsing in an unceremonious pile in the bottom of the boat. In the little wooden canoe alongside, Carlitos and his son were hysterical. It was the first time I had seen Carlitos laugh his lungs out. Gordon was almost choking to death, but it took me a little longer to see the funny side.

It is, of course, easy to be realistic after the event but I am now convinced it wouldn't – or couldn't – have done any real harm. A bite would have been extremely unpleasant, however, but there was no way a nine-foot anaconda could have swallowed my 190-pound frame – or was there? The most difficult part was having to get back into the water to get the shot – which I did about half an hour later. The story, at least, brightened up the crew's night aboard the *Natureza.*

The month of March passed quickly and we had some great images on film. It was almost time to return to the grime and crime of Manaus again and prepare for the next stage of filming – an expedition down the Amazon river to film dolphins underwater. Five weeks later we returned to the Xixuau, following a record fast turn around at Manaus to re-provision.

I had gone ashore to call home. After a trip away it's always a little nerve-wracking making the first call. I always dread the possibility of bad news. Frances, my mother's housekeeper, answered the phone. 'Hi, it's me, how's everything?'

'Oh dear, it's terrible, have you heard the news?'

My stomach came up into my mouth, 'No, what's happened – where's Mum?'

'Oh she's in the loo, no, it's Richard Dimbleby, he's died.'

I heard Mum's voice shout at Frances and the next minute she was on the line. It turned out that the previous night Frances had been watching a television programme called *Rock 'n' Roll Years* and during the show they had shown old black and white newsreel footage that included the announcement about the broadcaster Richard Dimbleby's death. Frances hadn't realised it was 30 years ago! Everything was normal at home.

In our short time away the water level in the Xixuau had risen dramatically by more than 21 feet. Much of the land had been transformed into *Igapo* – flooded forest. Every living thing – man, beast and plant – was to some extent affected by the change. Where Carlitos had hunted on foot in January, he now paddled to go fishing in May. Our filming target for the next two months was to capture this change in the landscape and film how it affected the lives of Carlitos and his family, and some of the creatures that shared their now-flooded world.

Several miles up the creek we came across the tip of a tree sticking out of the water. Only six feet protruded clear of the surface. In that small tangle of branches a colony of caciques were busy nesting. We turned our metal canoe into a floating hide by tying a film blind tent onto it. With this peculiar contraption I could float to within a few yards of the colony without the birds being concerned. It was thrilling to be so close and observe their behaviour. The females were busy knitting their sock-shaped nests together. All day long they tarried, flying to trees on the other bank, stripping thin lengths of plant fibre and returning to weave it into the hanging nests. The males simply stood on top of a clump of the nests and seemed to shout at them to hurry up. When a female returned to a nest to weave, he would ruffle his feathers and squawk noisily.

Caciques almost always build their colony close to, or around, a wasps' nest. This gives them quality protection, because if a predator looking for eggs or chicks gets too close the wasps launch a stinging attack and drive the offender away. We found this out one afternoon when I told Gordon I wanted to get even closer to the caciques' colony to see if this group employed such tactics. From the angle that I had been filming from for three days there was no evidence of a wasps' nest. We took the hide down and paddled in to the tree top. The birds scattered.

There was indeed a wasps' nest, ingeniously hidden among 15 or more of the birds' nests. It was the size of a melon and its outer surface was crawling with reddish coloured wasps. I was only an arm's length from it when I looked through the camera. Gordon steadied the boat by grabbing hold of a branch – but unfortunately that branch jiggled the wasps' nest. The camera was turning when something bounced off my forehead. Almost simultaneously Gordon screamed. His left hand clutched his left eye and his right arm paddled furiously to put some distance between us and the tree. He was clearly in a lot of pain. The wasp had ricocheted off me straight into his face and stung him on the eye-lid. Within five minutes his eye was closed.

Getting in closer to the caciques did pay off, if not for Gordon. I got the shots of the wasps' nest, and while we were paddling away to escape further attack I noticed an old bird's nest just below the surface of the water. We returned the next day with diving equipment to take a look. It was clear, dark water almost 30 feet deep. I sank down to the bottom imagining that two black caiman might be hidden in the tangle of submerged forest. Looking up from the bottom, the tree was beautifully lit with shafts of sunlight filtering through the still branches. I could clearly see the colony above the surface and Gordon peering down at me keeping a healthy distance from the wasps' nest. Several flooded nests swayed in the water as though pushed by a gentle breeze. It was apparent that the birds had built the colony lower down in the forest but had been caught out by unseasonal flooding. They had then moved to the very top of the tree where they now continued to rebuild.

'Only 228 shopping days left until Christmas,' Gordon announced at 7.30 in the morning on 9 May and then, leaning over from his tartan-patterned hammock, handed me a small gift-wrapped parcel – it was my birthday present. 'You get the other half this afternoon,' he said, not telling me what it was. At the dawn of my 42nd year, to the background of red howler monkeys calling nearby, I mused on how time flies. I looked into the mirror, splattered with dead mosquitoes, and plucked three ominous grey hairs that were trying to hide in a swathe of much better-looking black ones sprouting from my chest. I decided that I felt younger than my years. Then I shaved the 'salt and pepper' stubble from my face and resolved to accept growing older with grace. As I smeared on the shaving foam, I allowed myself only a brief thought about the family and friends I wished I could be with to celebrate, and the rack of lamb at Portofino's restaurant in Lytham St Annes that I would have ordered. I could almost smell the garlic and rosemary.

We never think about days off in this business as there is nearly always something to do. For me filming isn't a job but a way of life and, of course, I enjoy it immensely. Every day is different, every turn in a river or on a forest trail reveals a new view and possible surprises. I never take anything for granted and my 41st birthday was one of those days.

Gordon and I loaded the camera equipment into the boat and set off to do a day's work. First on the short list was, believe it or not, to find and film manatee excrement.

'I bet we are the only two people in the whole world looking for Amazonian manatee droppings today,' I said to Gordon.

As we pottered along a narrow creek, a troop of howler monkeys were giving it their all. It sounded as though one in particular was going for the 'howl of the year' trophy. Their guttural vocalisations are one of the strangest sounds of the rainforest. The sky was low and threatening rain as we floated through a pasture of red flower petals. Above our heads was the flowering tree that had exploded with blossoms overnight. Against the dark sky it was simply beautiful. We cut the engine to take a photograph and watch the dozens of humming birds pollinating the flowers. Euglossine bees, fluorescent green and bigger than grapes, zoomed in and out of the branches, pausing briefly to drink the nectar.

We eventually found several lumps of manatee droppings floating in the water and filmed them. As I packed the camera away into its rucksack, Gordon said to me, 'Why don't we go up to the giant otter den and see what they are at?' It was an indulgence, I knew, but what the heck, it is my birthday I thought. When we arrived there an hour later we literally bumped into the whole group. Five otters huffed and puffed for a minute only a few yards from our canoe, before getting back to the serious business of fishing. We pulled the boat into trees opposite the holt. There we sat and chatted quietly for three hours, watching a sloth pass overhead, listening to spider monkeys chattering behind us, and opening the other half of Gordon's birthday present to me. He had tucked a bottle of Tobermory malt whisky into one of the rucksack's pouches. A few nips with splashes of creek water went down 'singing hymns' – as we used to say in Lancashire.

The sun rose and set each day in resplendent skies and for the best part of a week every month we had such clear moonlit skies that we could venture up the creeks without using torches. Many nights we would just lie out on the upper deck of the *Natureza*, gazing at the stars, and talk about life. We were approaching a crossroads in our lives together. Gordon and I had been together for over five years and because we got on so well we could almost read each other's minds. As a cameraman I couldn't have had a better assistant. That I knew, but I also knew the time was coming when he would have to strike out on his own.

'How do you see our relationship?' I asked Gordon.

His reply summed up the joy of living and working with him. 'It's like marriage, but the sex is better,' he quipped. We decided that this film would be his last as my assistant and set about trying to plan his future.

Chapter Fourteen

Storm Clouds over the Amazon

In late March, while waiting for the waters to rise before continuing filming at the Xixuau, we left Carlitos's paradise to return to Manaus and prepare for the dolphin expedition.

The boto dolphins at the Xixuau kept mainly to the wider river channel in the Jauaperi, where it was difficult to film because the water was murky. INPA, the National Amazonian Research Institute, had agreed to allow its aquatic mammals expert, Fernando, to accompany us on the dolphin hunt. I liked Fernando, he always seemed orderly and sensible, and was obviously as honest as the day is long. He had told us of a place many miles down the Amazon, near the town of Santarem, where the Tapajos, a clear-water river, poured into the great Amazon. He had worked up the Tapajos with dolphins before and thought it was our best chance of getting underwater footage of them. We arranged to sail for Santarem on 14 April.

Manaus was being deluged daily with torrential rain, the stifling humidity bringing us out in a sweat despite there being no sun. At times the main roads resembled raging rivers, with Volkswagen Beetles being swept along as though on a revolving fairground ride. The regular flooding had one benefit, though, because it cleaned the streets of the daily garbage, piling it up into deep gutters where the vultures would reduce the mounds to almost nothing.

The week before our departure to the Tapajos involved the usual madness of trying to get all our equipment and supplies organised. The Tapajos was a long way away and the expedition required meticulous planning. We had decided to give up our apartment this week. We had had happy times there, but being on the top floor of an 18-storey block, there was nothing above our ceiling other than the flat roof and the rainy season. More than a third of the ceiling plaster had fallen in and most rooms leaked badly. After a surprisingly brief search we found a delightful rustic bungalow in a suburb of Tira Dentes, about five miles from Manaus centre. It was a big improvement. For one thing we were on the ground floor, and had access to a small garden with a swimming pool.

Two days before leaving Manaus I put a call through to Survival to let them know our immediate plans. Mike Linley came on the line, his voice sounding terrible. 'Nick, have you heard the awful news?' His tone frightened me. 'We received a call two days ago from Sumatra – Dieter has been killed.' I was stunned. Dieter Plage, our unit's universally respected and loved cameraman, had fallen from his airship in the Sumatran rainforest. Mary, his wife, had been there when it had happened. It was too dreadful to take in and I felt sick to my stomach. I could only imagine the reaction within Survival – complete and utter shock.

All of us take risks in this business, but Dieter's exploits were legendary. He would have been the first to put those risks into context. When you live and breathe wildlife films the dangers, although sometimes present, are always calculated. I was a relative newcomer to Survival, while Dieter had been with them for 25 years, and I was saddened that I wouldn't have the opportunity to get to know him better. On the few occasions when we did get together he always took the time to be interested in what I was doing. He was a most remarkable man and with his passing the world of wildlife film-making is an emptier place. This tragic news cast a shadow over our leaving.

By nine o'clock in the evening of the 14 April we were all aboard the *Natureza.* Four hundred litres of diesel oil were being siphoned into our fuel tank. The atmosphere was heavy with the smell of fuel. We were waiting for the moon to rise so that we could set sail for Santarem, 450 miles east of Manaus. The journey would take three days. We cast off, waving goodbye to Antonieta and Kesa, Fernando's wife.

A yellow moon shone above the tree line over the other side of the Rio Negro. Gordon and I seized our usual hammock positions. Slung under a small canopy on the upper deck, we could swing, read, rest or simply watch the scenery passing by for the duration of the voyage. Fifteen minutes out into the river I looked back at the port and city. It never ceased to amaze me how the water, the lights reflected in it, and the distance seemed to sterilise and decontaminate the place – even the Petrobras power station looked pretty at night from the river.

Apart from ourselves and the crew, we carried four additional passengers on this trip: Fernando and three other INPA functionaries, who were coming to assist in locating the dolphins. João Pena was about 45 and had a kind wrinkled face. He was unusually tall for a local man and had broad shoulders that gave him an air of being fit. His companion, Nildo, ten years younger, was short and tubby. His face, even when he wasn't smiling – which was rare – always looked happy. The third man, Severino, was the eldest, about 55, with a worn-out look. 'He likes to drink too much,' Fernando told me.

Then there was Donna Maria, our new cook. Donna Maria was large – very large, in fact. We met a lot of people in this region with eye problems, and Donna Maria's could only be described as wandering. Her first meal took us all by surprise – fish in Provençale sauce. It was so delicious that we praised our luck having such a good cook on board. The next morning we got boiled eggs – also in Provençale sauce. Unfortunately that was the only recipe Donna Maria seemed to have mastered because everything, from breakfast through lunch to dinner, every day, was done Provençale!

Sailing down the Amazon meant we needed a captain with experience of that route. Senhor Silva was short, bow-legged, but trim and fit-looking for his 60 years. There was an abruptness about his movements and speech that made me think he would probably be a bit of a handful at times. He was extremely polite and almost dipped his head when he shook our hands. 'Don't worry Sir, you are in safe hands!' he said to me in Portuguese.

'I wish he hadn't said that,' I muttered to Gordon, 'because now I am worried!'

'You could paddle a canoe through his legs without touching them,' Gordon replied.

At midnight we crossed the 'meeting of the waters' where the Rio Negro joins the Amazon. Our boat rolled and pitched across the lumpy expanse of water and, although the moon cast good light, we couldn't see the place where the 'black and white' waters met. The heavier than usual sway of the boat and melancholy thoughts about Dieter kept me awake.

Every so often I could hear chirping and peeping sounds. They would rise and fall, and after a few minutes fade into the night. Then a mass of floating vegetation came into the area illuminated by our boat's lights. It was some 15 feet long by about six feet wide, a mat of water hyacinth and coarse grass. We identified the calling sounds as coming from frogs hiding amongst the plants. Islands of vegetation were a common sight on the Amazon having broken away from the great floating meadows of the flood plains.

The following morning I awoke at six o'clock. My hammock swayed with the motion of the boat, as did Gordon's. Every so often the pitch of the boat would change and slam us into each other. I went up to the bow and passed Silva at the wheel. He was rubbing his tired eyes with his knuckles when he said, '*Bom dia, Senhor Nick.*' I looked forward and took in the awesome expanse of water. The *Natureza* was absolutely dwarfed by the river. It was more like an ocean and our tiny craft seemed so vulnerable bobbing along in the middle of it. The sun was rising fast into a clear sky, the distant forests on either bank mere thin green lines. Four tucuxi dolphins surfaced and dived in front of us, then a few seconds later one of them broke the surface again, spiralling into the air and somersaulting as though it was having the time of its life. At this time of year, in full flood, the water of the Amazon was rushing at great speed eastwards towards the Atlantic. You could feel it pushing the boat – we were in its grip.

The amount of tree debris was alarming. I asked Silva if it were wise to sail in the dark. '*Não tem problema,*' he replied as tree trunks overtook us, and we dodged water-borne debris that could have sent us to the bottom.

We dined on fish and rice that evening, watching the sun drop below the tree line. I asked Fernando to tell Silva that I would rather be a day late in getting to the Tapajos than risk disaster. He returned a few minutes later and said that Silva was happy to continue as long as there was a moon. No sooner had Fernando parked himself on the bench opposite me when the *Natureza* shuddered violently and slewed sideways, spilling our glasses of guava juice. We had hit a tree trunk.

Manuel jumped into the fast boat and, with Gordon's help, fixed a line to the *Natureza*'s bow. The fast boat pulled the *Natureza* slowly for 20 minutes until we reached the south river bank. There we managed to tie up to a tree and went below to assess the damage. It almost couldn't have been worse. The shock of the collision had broken the propeller and bent the shaft as well as a gearing plate. Silva seemed curiously quiet. Without any means of power until repairs could be done we were helpless. The nearest place to get the boat repaired was Santarem, two days' sail away.

A tow was out of the question. The few boats that were visible at dawn were miles away running along the centre current of the river. We set to fixing our fast boat to the stern of the *Natureza,* using a heavy chain to tie her to the rudder post and stern rails. With Silva steering and the outboard engine pushing us, we motored along at half normal speed, refuelling the engine every hour – but at least we were moving. We estimated we were eighteen hours from Santarem. At two o'clock Fernando and I were on the bow talking about dolphins and his hopes that INPA were going to put money into a new aquatic mammal centre. To our north the sky had turned black. Tropical storms are a common event here and with such vast expanses of water the violent winds that precede them can turn a placid surface into a maelstrom. Judging by the encroaching darkness this one was going to be bad.

'Better pull into the side and tie up for an hour,' I called out to Silva in Portuguese. He looked at the sky and replied, '*Não tem vento*' (no wind). Captain bloody Silva couldn't have been more wrong. A strong gust of wind hit our port side and Silva turned the *Natureza* a little to port, heading further towards the middle of the river. 'This is madness,' I said to Fernando. 'Tell the bugger to stop now!' It was too late. The surface of the Amazon whipped up quicker than I could have imagined. Three-foot waves pounded us. We were pitching violently. Silva panicked and started shouting like a maniac – at himself. The chains holding the fast boat to our stern snapped. We watched in terror as it bounced in a curve around our bow. Manuel, out of his brave mind, leapt into it from the bow post. I shall never know how he did it without killing himself.

We were about half a mile from the shore and side-on to the wind. Waves were coming over the lower deck into the saloon and Donna Maria was screaming. Twice I shouted to Gordon, 'This is it, we're going over,' but the *Natureza* half righted herself. I braced myself in the galley doorway, looking at £200,000-worth of filming gear, wondering how long it was going to take to replace. I will have to fly back to London to see the insurers – all this going through my mind in a flash. Fernando was trying to get a life jacket around Donna Maria. They looked as though they might have been about to dance, but Donna Maria was still screaming at the top of her lungs.

We all *knew* we were going to capsize and sink. The brutal gusts of wind pushed us closer to the shore, but the shore here was *várzea,* a flood plain, under six feet of water. There were several wooden shacks on stilts and small coppices of trees. I struggled up the stairway at 45 degrees and came face to face with Gordon. He had a life jacket on and was clutching a small Bible in his left hand. His face said that he thought we were goners. Just for a moment, seeing Gordon expecting us to go under, my spirits flagged and I was afraid. At that very second I remembered Dieter.

It was a submerged tree that had caused us to be in our present precarious situation, and it was another one that saved our bacon. We heeled over sharply again and for the fourth time we thought we were going to capsize. The hull struck another trunk which turned us stern to wind and brought us fully upright. We came to a halt in a copse of trees. Manuel managed to sling a rope over a thick bough and, as we all held it, the boat veered around to face the wind. The relief on board was audible – Donna Maria had stopped screaming.

Below decks, equipment cases were floating in 18 inches of water among tins of sweetcorn and soggy packets of cream crackers. My precious few PG Tips tea-bags were no more. Our saving grace – the second tree trunk – had been at a price, though. The water level inside the boat was rising. Two bilge pumps were switched on and Gordon, Manuel and I squeezed into the engine compartment on all fours. Two plumes of river water were spraying the underside of the lower decking. Using bits of ripped T-shirts Manuel pounded them into the split timber with a heavy chisel and hammer. It slowed the ingress to a trickle.

The storm disappeared as quickly as it had appeared and by now most of us were laughing, from pure relief. But Donna Maria wouldn't remove her life jacket even though the danger had passed. I looked into Silva's face and said, '*Não tem vento neh?*' (No wind, eh?) He didn't smile and his hands were trembling.

Astonishingly the only casualty was the second fast boat and its outboard engine, which had sunk. But all our camera equipment was saved by the sturdy waterproof Pelican cases.

The following day in Santarem the River Police said that only a few miles west of us another boat hadn't been so fortunate, it had gone down taking five men with her. I should perhaps have paid more attention to Bates' experience when he was in this locality:

'...a black cloud arose suddenly in the north-east. João da Cunha ordered all sails to be taken in, and immediately afterwards a furious squall burst forth, tearing the waters into foam, and producing a frightful uproar in the neighbouring forests. A drenching rain followed: but in half an hour all was again calm...'

Despite the tiresome wait for repairs to be completed – it was going to take two days – the area around us was well worth exploring and Gordon and I went to investigate in the metal canoe.

During the months when the water is high, many areas along the banks of the Amazon river flood. The clouds of mud suspended in the Amazon river giving it the distinctive 'coffee with milk' colour are rich sediments that originate thousands of miles away in the Andes mountains. The sediments fertilise the *várzea* as the turbulent river silts settle and help to make them one of the most productive fish areas in the Amazon. What appear to be lush grasslands are in fact floating meadows.

The tallest grasses reach six feet and mixed with them are water hyacinths with delicate pinkish red flowers, water lettuce – a favourite food of the Amazonian manatee – and carpets of lush ferns. Many small communities scratch a living in these flood plains. They have to adapt between a dry season, where they can till fertile land, and a wet season, when their feet don't touch land for six months. The daily existence of people and animals revolves around walkways, houses, farms and even churches supported on stilts which generally, but not always, keep them just above the surface of the water.

One community we visited was Aracampina where 386 people lived, 66 families in all. Most of the houses were connected by frail looking, narrow planked walkways suspended just a few inches above the muddy coloured water. Behind them old unserviceable wooden canoes had been mounted on stilts and were used as gardens, overflowing with medicinal plants and herbs. Small children played, dogs crapped, ducks waddled and pigs meandered on this amazing network. Several of the houses had stalls raised less than 18 inches above the water with cattle crowded into them.

Early the following morning we motored across one of the large *várzea* lakes. The sun was rising quickly and the surface of the lake shimmered like glass. The surrounding meadows were a fresh lush green and almost felt like the English Lake District on what would promise to be a hot summer's day, except that the temperatures here in an hour's time would make the record books if in Cumbria.

We noticed four strange dark humps coming towards us on the lake a few hundred yards away. 'What the hell are they?' Gordon asked. They turned out to be canoes, covered in a mountain of grasses. At the front of each floating hay-stack a man, face shadowed by a straw hat, paddled. They had been cutting the aquatic grasses, which they use as fodder. We followed one of the men to his home. As he paddled into a small channel off the lake we could see the back of his house and the cattle pen. The animals, like on any farm, knew it was feeding time. They were penned into an extremely small area and began to jostle each other as bundles of the grass were thrown over to them.

I mused that these houses must be stronger than they looked if they could survive storms like the previous day's. We passed a crudely built wooden shack, connected to the main house by a single eight-inch by 15-feet long plank of wood. It was a toilet. Outside the shack door, large slabs of caiman meat hung from a wooden rail, drying in the sun. A cloud of flies rose into the air from the strong smelling flesh as we edged past, holding our breath, in our metal canoe.

The repairs completed, we were to set out the next morning to find and film the river dolphins in the Tapajos. It was going to be smooth sailing from now on, I tried to convince myself. In an effort to revive our jaded palates, I suggested to Gordon that he talked to Donna Maria – subtly, of course – because we didn't want to hurt her feelings.

'To encourage her away from things Provençale, give her our tin of prized and expensive Danish ham – it's worth sacrificing,' I urged him. He disappeared below to test his diplomatic skills.

At 6.30 pm, with a certain air of spreading excitement, we all gathered in the lower deck dining area. Plates and cutlery had already been laid out. 'Grilled with fried eggs would have been OK for me,' I said as the steaming pot of rice was laid before us.

'What's she done tonight?' Fernando asked Gordon.

'Er, well, I told her how to to er … grill it – you know,' he replied without conviction.

'*A junta está na mesa*' (dinner is on the table), Donna Maria announced as her massive frame brought another, smaller, pot through. She about-turned and, singing to herself, trotted heavily down the steps.

Gordon lifted the lid and a rush of steam billowed out. He peered in, wafting the clouds of vapour with his hand. 'Oh bloody hell – look,' he invited. Donna Maria had simply cubed £15-worth of prime Danish ham and mixed it with a sauce of onion, tomato paste, and garlic! Fernando and Nildo were chuckling loudly when Donna returned with a flagon of fruit juice.

'*Tem um problema?*' she enquired of Nildo, looking like she might give him a guava juice shower at any moment.

'*Não ... não, tudo está perfeito, Donna Maria*,' Nildo replied like a little boy who had been chastised by the headmistress. The Provençale ham-hash tasted just fine.

I remembered that Henry Walter Bates left Santarem and entered the Rio Tapajos on 8 June, 1852, exactly 140 years, 9 months and 21 days before we did the same thing. I had had plenty of time to read these past few days and trivial facts like that interested me:

> '...it is about a 1,000 miles in length, and flows from south to north; in magnitude it stands the sixth amongst the tributaries of the Amazons. It is navigable, however, by sailing vessels only for about 160 miles above Santarem... Our course lay due west for about 20 miles. The wind increased as we neared Point Cururu, where the river bends from its northern course. A vast expanse of water here stretches to the west and south, and the waves, with a strong breeze, run very high. As we were doubling the Point, the cable which held our montaria in tow astern, parted, and in endeavouring to recover the boat, without which we knew it would be difficult to get ashore on many parts of the coast, we were very near capsizing. We tried to tack down the river; a vain attempt with a strong breeze and no current. Our ropes snapped, the sails flew to rags, and the vessel, which we now found was deficient in ballast, heeled over frightfully.'

We followed Bates' ancient route up the Tapajos to a small village called Altar do Chão – Altar of the Earth. We had seen many boto dolphins on our voyage. They were tantalisingly close, but we were faced with an irrefutable fact – the water level was too high for us to run out INPA's nets to contain them in the area for an hour or two, long enough to get below and film them. Our one chance was to find a local fisherman to lead us to parts of the river or creeks where the dolphins were known to congregate and where we could use the nets. A local person in Santarem had already told us that the man we needed was Vincente. However, like everything out here, finding him was not that simple.

'He's a wanted man,' Fernando told me after speaking to a stern-faced river police. We assured the Capitania that we would forget about looking for Vincente and set off for the other side of the river to the Igarapé Jari. During our wait for the repairs on the *Natureza* to be completed, we had met a family who lived in this creek and they had opened their front door to us without reservation. We decided to go and ask if they could help us find Vincente.

Twenty-five families lived along the Jari. At this time of the year, the high water spread well beyond the usual limit of the muddy creek bank far into the forests behind. Flocks of purple-black ani birds followed each other silently from one tree to the next to keep ahead of us. The *Natureza* almost filled the width of the creek and its upper deck brushed the branches of overhanging trees on both sides. At a fork in the waterway we set off in the metal canoe to Senhor Sebastian's house. Together with his wife and seven children, Sebastian was sitting on their front porch with two dogs and a small parrot. The water was inches below their walkboards. One of his sons, 12-year-old Chico, dashed along the planks and jumped into our fast boat before we even had time to tie it up. He sat behind the steering wheel, his face shining with delight, and pretended to drive it, making zooming noises while smiling at us. He had never been in a boat with a motor on it before.

Fine rain began to fall. I asked Sebastian what he was going to do if the water came up any higher. His toothless wife laughed as her husband predicted that the water had already stopped rising this year. The baby in her arms splattered diarrhoea over her frayed dress and right leg. She laughed again and, dipping her hand into the water, splashed herself until she was more or less clean. Then she dunked the naked baby twice, grasping both of its arms over its head. It didn't make a sound. Sebastian did know

Vincente, and explained how we could find him. Before we left, Gordon and I took Chico for a spin down the creek in the fast boat. It was the thrill of his little life.

The man that we – and the Capitania – were after lived on the other side of the Amazon river about three hours away. Sebastian told us of a lake close by where there were many botos dolphins and suggested we went there first. We passed half a dozen fishermen in canoes as we entered the lake and they all avoided us as though we carried an infectious disease. João Pena and Nildo tried to speak to them but they pointedly ignored them too.

The water was unruffled except where the dolphins surfaced. Fernando, Gordon and I counted more than 20. I scanned the strait to our left where it met with the Tapajos and I noticed three small wooden canoes with their 'silent' fisherman pulling themselves into thick reeds at the neck of the channel. João Pena said '*bombeadores!*' Before he or Fernando could explain, a fountain of water rose into the air and a second later the sound of a depth charge going off reached our ears. These men were catching fish using home-made underwater bombs. It's a dangerous and illegal practice and no doubt was the reason for the local fishermen avoiding us. With '*gringos*' on board we looked too official.

'Will you look at that?' Gordon cried out.

In the entrance to the lake, something redolent of a tidal bore was bearing down the channel towards the main river. It was a flotilla of more than 40 dolphins passing directly off our stern. It was an extraordinary sight. The dolphins had learned quickly that the underwater explosions meant easy pickings and dozens of dolphins were diving and surfacing, grabbing the stunned fish. Coming out of the reeds towards them were angry fishermen. As the first of their canoes reached the spot where fish were floating to the surface, the dolphins immediately disappeared, but occasionally a fisherman would launch an *azagaia* (a fishing spear) as he glimpsed one below the surface. Within a few minutes the dolphins passed our stern again and returned to the lake, presumably to await the next bomb.

These men were fishing to supply the markets in Santarem. It was just another place where man was in conflict with an animal for commercial reasons. We could not film the dolphins here because apart from potential problems with bombers there was no part of the channel we could section off without endangering the dolphins. We all agreed that Vincente, despite our not knowing what the authorities wanted him for, was the man we wanted.

At 11 o'clock on 28 April, Fernando and I left the *Natureza* on our powerful fast boat in search of Vincente. The weather was good as we set off, the sky a deep blue and the water, if not exactly mirror calm, smooth and undulating. A giant Amazon kingfisher flew by, clacking noisily, with a small fish in its beak. It took us 30 minutes to reach the other side of the river. There we turned south, as directed by Sebastian, and began to look for the entrance to a *paraná*, a parallel channel to the main river, called Surubim Açu. Fernando spotted a boat almost hidden in dense reeds and we nosed into the vegetation. The boat was a *batelão*, a low single-decked fishing houseboat, and we had found Vincente – as simple as that!

It looked like he was in hiding. The old boat's sides were shrouded in pieces of tatty blue tarpaulins and, at first sight, there appeared to be no one on board except for a scraggy muscovy duck perched on the low roof. Fernando clapped his hands and shouted '*Bom dia.*' No reply. We fastened our bow rope to the side of the fishing boat and Fernando lifted a piece of the tarpaulin. Inside the dark interior were six or seven people all lying in hammocks. A fat male face looked at us and grunted. Fernando asked him if he knew Vincente. '*Porque?*' Why – was the response. Fernando explained that we were looking for Vincente to help us to find dolphins for filming. Another hammock stirred and a young man, about 25, who looked like he hadn't slept for a month said something to Fernando in a thick local accent.

'It's him,' said Fernando.

As Fernando talked to Vincente I kept hearing the word '*dinheiros*' – money – and I realised it was a lost cause. A third man lurched from his hammock and swigged a bottle of *cachaça*. Everything about this bunch smacked of trouble. Vincente could organise dolphins for us to film but it was going to cost – a lot! Fernando didn't like the look of the man and said that he seemed 'illegal'.

Realising that this was the end of a costly expedition I said, 'Let's cut our losses, I don't want to waste any more time on this.' We untied our boat and pulled ourselves back through the floating grass to the channel. I felt the weight of the world upon my shoulders and wished I could have been back at home doing something mundane like watching *Coronation Street*.

Sometimes you make a journey in one direction but the return, even though the same route, seems wholly different. 'Is that the way we came?' I asked Fernando, pointing to the island as he strapped a life jacket on. The sky had darkened, the wind freshened and rain had begun to fall as we edged out into the main river – the Amazon. Between us and the island was a sea I had no wish to cross. Around us it was only choppy but in the middle the waves looked huge and powerful. 'Let's wait for the squall to pass,' I urged, and Fernando agreed. We motored into the nearest flooded tree line and fastened a rope to a branch. There we waited – and waited. Half an hour later, deep in ethereal conversation about how life passes one by so quickly, I looked out at the river and thought it had calmed a little. 'What do you think, Nando?' I asked. 'Shall we go for it?'

'You're the captain, you decide,' was his cop-out – rightly so.

Since I was a young and immortal 19-year-old I have had plenty of experience at sea with boats. I should have known better. The choppy waves metamorphosed into full-blown breakers and we pounded from one to the next. As we rounded the western edge of the island and I could see the final expanse of a mile and a half to the other side, I remember thinking we should turn back – but I didn't.

We slammed into the main current of the river, dipping into troughs bigger than our 15-foot boat, rising the next crest at an alarming angle. I knew that to try and turn the small boat around would have been the end of us and we had no alternative but to plough on. Our metal boat had no flotation chambers and if she flipped over she would sink immediately. Fernando's face was full of panic: both of his hands, knuckles white, were gripping the metal gunwale. Clutching the steering wheel, mine must have looked similar.

Somewhere in the middle of the horror I said aloud, 'Safety is only the distance between the north and south piers at Blackpool' (a stretch of, at times, treacherous sea that I knew well). I was speaking to myself, looking for courage. Water poured into the boat. There was no possibility of bailing it out as we needed both hands to hold on for dear life. Then, suddenly, before we could register this, we were in more even-tempered water. Fernando spoke, for the first time in 20 minutes. Laughing aloud he said, '*Meu cu está apertado*,' without a trace of embarrassment. He meant that his bottom was twitching! He then asked, 'What is Blackpool?' Safely aboard the *Natureza* at last, we decided to cut our losses and leave the Tapajos. We thought we might have better success in the Jauaperi next January when the water would be low and clearer.

On 30 April at 8.30 pm we steamed into the small Amazon riverside town of Obydos. It was a calm and pleasant place, without the usual hustle and bustle. Twenty or so small trading boats were tied up for the night, their low concave roofs stacked high with piles of bananas, coconuts and cassava roots. Five two-wheeled carts stood by, mules harnessed to them and drivers asleep in the traps, ready to carry cargo ashore. A steam whistle shattered the calm for ten seconds, announcing the start of the night-shift at the riverside Brazil-nut factory. On the promenade in front of our boat there was a bicycle contraption with a modified front wheel where a drive belt ran up to a contrivance over the handle bars. Pieces of sugar cane were dropped into it and a small boy, pedalling away, crushed the sticks. Other young street boys were moving from boat to boat selling cups of the *caldo de cana* – the juice extracted from the crushed sugar cane. It tasted horrible.

The following morning, before our departure, I wandered the narrow streets watching people prepare for the day's work. In the cramped doorway of one shop baskets of tree bark, plants, roots and other strange-looking objects spilled out across the uneven ground outside. Inside the shop was like an Aladdin's cave. Shelved from floor to ceiling and only ten feet wide, the place was filled with hundreds of glass jars containing pickled 'things'. I was standing in Senhor Bermuda's regional medicine shop. Here I met one of the most sociable present-day inhabitants of Obydos. Had he been there in Bates' day he would surely have been immortalised in his classic book. He spotted the *gringo* immediately and was at my elbow in an instant.

'*Quero atrair uma mulher?*' was his opening gambit (do you want to attract a woman?). Holding some ghastly-looking thing in a jar inches from my face, he rotated it so that I could see it in all its gory glory. It was a boto dolphin's penis.

Senhor Bermuda reminded me instantly of Captain Cockroach and I remembered Cockroach and his dolphin penis tale. Seeing me shake my head this short old man, with the most lived-in face I have ever seen, took me along the shelves proudly exhibiting horror after horror. Dead snakes and insects of every kind stared back. If it was an animal and male, he had its penis pickled somewhere. Those of caiman, coatimundi, tapir (looked lethal), monkey, paca, agouti, all floated in clear preservative. Then there were the eye-balls. The paca's is apparently good for gout.

Senhor Bermuda grabbed a dried, rotten-looking banana from a basket at our feet and almost shoved it up my nose. '*Banana regional*,' he exclaimed, smiling.

'What's it good for?' I asked.

'*Cura tudo*,' cures everything, he assured me. His cramped shop was filling with customers and Bermuda would shout to draw their attention to me. I seemed to represent some sort of status symbol. If only he had known what I really thought of the place, he might have stabbed me with the dried tapir's dick he was trying to sell to a woman standing next to us. In great detail he told me how to prepare a tapir's toe nail to cure labour pains and a dolphin's eyeball so that my own eyesight would improve. I told him who I was, not that it meant anything to him, but he nodded approvingly. I asked if I could interview him on camera about his trade. He immediately announced to the shop that he was going to be on television and then turning to me – with almost a theatrical aside – told me it would cost $50!

Half an hour later, with a tame capuchin monkey trying to make love to the furry microphone cover Gordon was holding, a midas tamarin playing hide and seek in my camera bag, and Fernando seated next to Senhor Bermuda, we began what is probably the strangest interview I have ever filmed. With all his interesting wares spread across three wicker table tops, Bermuda launched into what you could do with this and that. For the sake of continuity we had to film several things over and over again. It was stifling hot and there were many interruptions.

The final question was direct: 'It's illegal to kill these animals and sell their parts. Why do you do it?'

Sharp as a knife he replied, 'People want to buy – so I sell.'

One of my many faults is that I get intense when I am filming people and unreasonably impatient if things are not going right while I am looking through camera. As the sound tapes were being replayed I cringed with embarrassment as I heard my irritable voice uttering, 'Tell him to pick his penis up, sorry I mean the dolphin's. OK, now put it down. No no, he's not handling it properly, he keeps moving it up and down and I can't see it clearly… OK now tell him to hold it between his thumb and forefinger'… and so it went on.

Apart from the dolphin expedition, everything went surprisingly well until 20 August, 1993. That was the day we returned to Manaus to stock up with provisions for our final expedition to the Xixuau and the Rio Branco (to film a bizarre looking terrapin called *mata mata,* and to see what was happening with the jabiru storks).

Back in Manaus I met up with Marc van Roosmalen who asked me if I would be interested in making a film about marmosets, the smallest monkeys in the world. He wanted to take me to an area south of the Amazon river where three species of marmosets lived, which had never been filmed or studied. He told me the Sataré Indian women kept them on their heads for the tiny monkeys to eat their lice. I was hooked on the idea and we agreed to make a three-week expedition before December to research the story.

Marc also said he had found an area of rainforest 18 miles north of Manaus where we could set up a permanent base. As my mind overflowed with happy thoughts of a permanent forest home, I made my usual telephone call home and received dreadful news. My mother had been taken ill. Tests had shown that one of her heart valves was malfunctioning and she needed major surgery. My brother Richard lives in the United States and my sister Gillian in Canada, so Mum was on her own. The specialist at Victoria Hospital in Blackpool had told her that the operation would have to be done sooner rather than later and so I flew home to the UK the following week.

Mum seemed to have aged 20 years since I had last seen her and could hardly walk a few paces without having to sit down. First stop was the heart surgeon in London. We sat quietly in his library waiting room with two other men, who were dressed from head to foot in white robes. 'This is going to cost a packet,' Mum said in a hushed tone. We were ushered in and learned that the test results showed that she needed a replacement valve and possibly a single bypass. Mum was silent.

'What are the options for my mother?' I asked the consultant.

'If she doesn't have the operation, about 12 months. If she does, she will lead a normal life again. There's a two per cent fatality chance during the operation from a stroke.' Mum cut in, 'Right that decides it, when do you want to operate?'

'The week after next, in Surrey.'

I could tell that Mum was trying to take in the enormity of it all.

Ten days later I was in a private room in a Surrey hospital saying goodnight to her. Mr Parker would operate at eight o'clock the following morning and I wondered if it might be the last time I would see her alive. The hospital was staffed and run by nuns who floated about giving the place an ethereal atmosphere. 'I'll be here tomorrow when you come out of the operation,' I said, trying to control my emotions. There was a clear starlit sky and I thought of the Xixuau light years away.

The operation was a total success and at one o'clock the next day I was in the post-recovery room looking down at Mum. There were more tubes coming out and going into her than Kings Cross station. I stroked her forehead and to my amazement she half opened her eyes. Through the plastic mask she said, quite lucidly, 'Hello darling. I've just come back from Harvey Nichols, got some nice shoes,' and then she closed her eyes again.

'She's obviously been having a better time than me,' I half choked with relief to the nurse.

One week later I was driving Mum back up the motorway to Lancashire. Her recovery from such a major operation was nothing short of spectacular. I visited her for a couple of hours every day and for the rest of the time indulged in the luxury of the hotel where I was staying. I relaxed with books and magazines, nipped down for a dip in the pool a couple of times a day, and ate far too much, especially at breakfast. When you have spent years in the forest, there's little to beat a plate of cholesterol to get the day going. Three weeks later my sister arrived from Canada to take over the convalescent watch and I was back on a Varig Airlines flight bound for the Amazon once more.

Chapter Fifteen

It's a Wrap!

I arrived back in Manaus in September to embark on our final four months' filming to complete *Creatures of the Magic Water.* We had to revisit both the Rio Branco and the Xixuau. First I met Marc to find out the latest news about the planned marmoset expedition. Arrangements were well in hand. Marc had found an area of forest near Manaus that might make a base for the film, and a new home for me.

Igarapé Jacaré (Alligator Creek) is hidden from the world about 20 miles by river from Manaus. At high water, with a fast boat, it only takes an hour to reach the site, but at low water the journey can take up to another four hours, and only the determined will find it. Once off the main river the narrow creek twists and turns through forest that is flooded for half the year. The water was crystal clear – good for drinking and filming – and in many places we could see white sand at the bottom. Huge aldina trees towered above us and dozens of their large bum-shaped fruits floated on the mirror-calm water. Lianas sagged under the weight of enormous spiky bromeliads whose bright red flowers looked stunning caught in the sunlight. Palm fronds draped down to the water and scarlet red dragonflies rested on them. Humming birds hovered just in front of our faces as we paddled up the creek and occasionally a grey piha bird screamed, its evocative call echoing through the rainforest. This was the perfect place, I thought, close enough to Manaus for provisions, yet almost impossible for anyone to find.

Several miles up the Jacaré, Senhor João had lived and farmed a small area of forest for many years. He was now intending to move much further up the creek to make a new roça (or farm) and it was his old place that we were interested in taking over. João's dark wrinkled face smiled a lot as Marc talked to him about the arrangement. Nailed to a tree a few paces away from us was the body of an ornate hawk eagle which he said he had shot because it had been after his chickens. Although the Caboclos can – and do – just move in and settle land virtually anywhere, it is a different matter for foreigners. There was no such thing as a title for the land, and so we needed to agree a price to allow us to live there.

It turned out that João was a good carpenter and part of the bargain was that he and a friend called Moses would build the house to my design. Only a few hundred yards from the old shack the forest was pristine and would be perfect for building filming towers. Marc, an expert neotropical botanist, explained to me that many of the larger trees here were the right kind to attract monkeys when they were in fruit. The weird names of the trees rattled off Marc's tongue so easily, but at this stage I couldn't tell an *Mimosaceae* from a *Sapotaceae* – or even pronounce them properly.

As we explored the forest I noticed something move on the floor about 12 feet from me and shouted to Marc to come and look. It was a large, brilliant green six-foot long snake and was coming straight for me at great speed. As I moved away, it followed, hissing and spitting and waving its head from side to side. It was a remarkable display. Although not a venomous snake it is alarming to have such a creature coming at you when you are used to seeing them flee. Marc cut a sapling and danced around the lungeing snake, before pinning it with the stick so that he could pick it up. Wrapping itself around his forearm and waist it inflated its entire body with rage and the last six inches of its tail shook like a rattlesnake's. When Marc let it go we had to move away quickly as once again it came slithering after us. It eventually gave up the chase.

Back in the old farm clearing we sketched a plan of how I would like the house to be built. It had to be spacious and practical, almost completely open-plan and as comfortable as possible. Any fool can make things hard – that's easy! But I was determined to make life as comfortable as would be practical in this rainforest; after all I might be staying for several years. We couldn't do anything about the temperature and humidity, but I could see how, with a little careful thought, things like a cesspit and sit-down toilet could be built and a water tower set up to supply the house with fresh creek water. It was to be a real home rather than a temporary camp.

Most of the building materials would have to be brought in by canoe, and Marc, who was keen to get the project on the move, would supervise the building while I finished filming up in the Rio Branco and Xixuau. When we shook hands with João he told us it would be ready in seven weeks! So the next time I would return here I would have a new home – it was a thrilling notion. We disturbed a resting southern river otter, very similar in appearance to the European otter, on our way out of the Jacaré and I took this as a good omen. Resident otters must mean good fish stocks.

We intended to leave Manaus for the northern territories at nine o'clock in the evening on 24 September. Unfortunately, I had paid a small advance to Carlinho, Carlitos's eldest son, when he had asked me earlier in the evening for a sub to leave with his girlfriend. With a glint in his eye he had promised to be back before 8 pm – it was now almost 11 pm. We had employed Carlinho as our *prático* for this trip because he knew the route to the Xixuau blindfolded. Gordon set off to try and find him.

While the rest of us waited in the crammed and stinking waters of Educanos – a port overspill creek – I talked to the old man sitting on our boat's bow. His name was Papagai (parrot), and he lived on the *Natureza* as security guard when she was not rented out. It was clear why he was called parrot – he had a nose like a macaw's beak and the shiny curved end of it almost touched his bottom lip. He was barefoot and wearing only a threadbare pair of corduroy pants held up by a belt with an Elvis Presley buckle. The skin covering his bony frame looked like tanned rawhide. I was leaning against the boat rail facing Papagai's wizened toothless face as he counted off the names of the monkeys that he liked to eat when, without any warning, he jerked up the shotgun which had been resting on his lap and fired it into the night about an arm's length from my right ear. About 150 feet away on the shore a dark prone figure dragged itself along through putrid garbage and disappeared between two shacks. '*Ladrão*,' thief, Papagai muttered. I must have been still in shock at having had 50 or so lead ball-bearings whiz past my right ear at close quarters because my hands were shaking. The man on the next boat laughed with Papagai about what had happened. Five weeks later he was shot dead on his own boat by a nighttime intruder.

Shortly after midnight, while thunder rumbled overhead, Gordon and Carlinho arrived back together. Gordon had been directed to a motel close to the house where Carlinho lodged and there he had found him, spending the 'loan' intended for his girlfriend. Unrepentant and with the cheeky grin he so often wore, Carlinho cockily slapped me on the back as he walked past me towards the helm. Somehow I couldn't help but like him.

We pulled out of the creeks relieved to breathe in air polluted with something other than human excrement and tied the boat up to one of the enormous floating fuel stations to take on diesel and oil. The floating station is simply a huge tank containing thousands of gallons of fuel and it reeked of petroleum. In the middle of a mountain of silver-coloured propane gas bottles and five old fuel pumps a man stood with a lit cigarette in his mouth. I imagined the spectacular display this place would create as it exploded into the dark night sky. I walked into the shabby office to pay only to find a fat man slouched at an old desk, hands folded on top of his large bare belly, feet propped up on top of the desk, watching football on a small black and white television. In one corner of the desk, less than three feet from his feet, was an in-tray overflowing with oil-smeared invoices. Curled up on top of the invoices – fast asleep – was a huge brown rat.

Our first destination was a lake far up the Rio Branco called Lago do Mata Mata, where we hoped to come across those bizarre-looking terrapins after which the lake is named – the *mata mata*. Our cook Donna Maria had disappeared without warning and Gordon said he was worried because her replacement looked remarkably normal – which it turned out she was. Manuel's first mate, Cabeça, was almost six feet tall and had the physique of a body builder but his head was huge, hence his nick-name.

I read four books over the three days it took to reach our location. Over the years I have been in Amazonas books have become an important source of escape for me, one of the few ways I find I can take my mind completely off work. Apart from the many that I read for research, mainly scientific, I have a library of fiction and non-fiction that one day I shall have to send home by ship. I particularly like reading about the escapades of other travellers, because I identify with their experiences, especially with those who remain isolated from civilisation for long periods.

We passed the IBAMA post in the mouth of the Rio Branco to find that nothing had changed since my last visit. Chico Preto was still in his chair, mangy dog at his feet. I asked him if he ever got lonely out here. '*Acostumado*,' he replied with a gentle smile on his face. He was obviously happy with his own company – which was just as well. Two days later we called into Aricura where we had filmed the *expansa* terrapins months before. As our boat pulled in to the river bank Luiz was standing at the top of the ladder, his left hand swathed in a bloodied rag. Rooting in the earth next to his feet was Gretchin – a tame peccary. While trying to remove an awkward hook from the mouth of a large black piranha the fish had lunged and grabbed his index finger. It had almost severed his finger at the first knuckle. In a normal world he would have been taken to hospital – here we just delved into our medical kit. I produced the suture needles – one of the essential skills I had picked up! – but Luiz frantically waved his good finger in a 'no no' sign. Instead I just cleaned and dressed the wound, applying lashings of antiseptic cream.

Luiz had some bad news for us. A team of fishermen had passed through the area four weeks earlier and camped on the sandbar in front of the samauma tree where the jabiru storks had started to refresh their nest. Luiz found the remains of an adult *expansa* terrapin in the embers of the fishermen's fire, and hadn't seen the two birds since. It sounded grim. We could only take another look and if nothing was doing, carry all three tons of the scaffolding back to the boat for the return to Manaus.

A few hours later we left Luiz, giving him some packets of rice, coffee and beans and fresh dressings for his piranha bite. He gave us instructions to find Mata Mata lake, about half a day's journey north. He used to fish there, but on his last visit an 'evil spirit' had 'chased him' to the mouth of the lake.

'*Melhor não ir lá*,' (Better don't go there!) Luiz advised as Gordon and I waved goodbye. But we hadn't sailed for five days to turn round now. The only evil spirit Luiz would have encountered was *cachaça*.

The entrance to the lake was almost blocked by a sandbar but water 18 inches deep would allow us to use our metal canoe. White sand, like a deserted Caribbean beach, swept out from the forest. At the water's edge and wallowing in the cool river were a herd of capybara, the largest rodents on earth.

'Bloody hell, look at them, Nick,' Gordon yelled. In the shallows and straggled along the shoreline were more than 20 of the gigantic creatures. The ones in the water only just had their heads clear of the surface while the rest remained motionless on the sand. Like otters and alligators, their ears, eyes and nostrils are aligned on the same plane so that they can remain almost invisible in the water. But it was their hugeness that amazed me – roughly the size of a wild boar. As we approached they bolted for the forest. Carlinho asked me if we could pull into the sandbank so that he could collect their '*cocó*'. Living in Amazonas for so many years has taught me to expect the unexpected, but why anyone should want to collect the droppings of the world's biggest rodent threw me. Capybara dung, he said, makes an excellent medicinal tea, '*Muito bem para seu estomago*' (very good for your stomach).

We tied the *Natureza* to a tree and climbed into our fast boat. The channel was only ten feet wide and extremely shallow, too shallow for the small outboard motor. As we paddled through the narrows, the bottom clearly visible, a mountain of water suddenly erupted in front of us as a pink river dolphin panicked and rocketed past us for the deeper river channel. It opened out into the lake and our first sight was a big splash as a giant otter flopped off a fallen bough into the water. The water was a dark muddy green and Amazon kingfishers clacked noisily around the edges. Herons and anhingers took flight at our intrusion and the surface rippled with dozens of fish.

Wearing hats and sunglasses to protect us from the searing white light of the midday sun we paddled into the middle of the lake with Gordon perched on the front of the metal canoe, rough wooden paddle in his hands. Without any apparent reason he suddenly fell, fully clothed, into the water. Immediately a large object was thrust up and over the edge of the boat. It was a *mata mata* terrapin. Carlinho grabbed the prehistoric looking beast with both hands and put it in the bottom of the boat, upside down, as Gordon heaved himself aboard. I told Carlinho to turn the terrapin over so that it was the right way up. He laughed as he did what I asked because, as he knew, the creature immediately tried to escape, causing chaos. It snagged our equipment bags, almost dragging one of the film containers overboard. Without another word of interference from me Carlinho retrieved the bag and placed the mata mata upside down again, where it remained motionless.

Lago do Mata Mata was the place the terrapins came to lay their eggs and there were dozens of them in the lake. For some reason they float on the surface for several minutes at a time. Perhaps this is to absorb warmth, because unlike others of their kind they do not come out of the water to sunbathe. The only time they emerge onto land is to make a nest and lay their eggs. They are truly grotesque looking creatures. Their large shells are ridged like rocks – good camouflage on the bottom of a river – and a strange head protrudes with a ridiculous pointed upturned nose like some creature from *The Wizard of Oz.* They have many strange fringes of skin on their faces that apparently detect the movements of the small fish on which they feed. When upset, we soon discovered, they produce an obnoxious stink and for this reason almost nobody bothers to eat them.

We pulled into the side of the lake to let the terrapin go, filming it as it stirred up the wet earth with its feet and scrabbled down the muddy bank. Before the water had settled Manuel shouted something in Portuguese and pointed out to the middle of the lake. A fishing hawk was hovering about 30 feet above the surface. We all watched as it plunged into the water and grabbed a big fish. It struggled for a few seconds to rise clear of the water. The fish was thrashing violently, but the bird's large claws held it fast as the hawk flew to a nearby tree. There it started to tear its catch to pieces. As we made our way back out to the main river an armour-plated armadillo about the size of a Scottie dog galumphed down the river bank and dived into the water. It then porpoised its way across the lake and ran up the far bank, oblivious to our presence. 'What a fabulous place,' I said to Gordon.

That evening as we discussed the plan for the following week, we played gin-rummy while relaxing in our hammocks. If, indeed, the jabiru storks had disappeared from their samauma tree we would load the scaffolding on board, which would take a day and a half, and set off for our final visit to the Xixuau to film Brazil nuts and water vines.

After cleaning our teeth we drifted off to sleep to the gentle strains of '*When the moon hits your eye like a big pizza pie – that's amore*' by Dean Martin. It must have been the prompt for my dream about entering a competition to find who could eat the biggest deep-pan pizza. Mine was overflowing with garlic, salami and artichokes that night. Gordon remembered his dream, too — he had woken up one morning to find that he had grown a second penis!

By six the next morning we were sailing south on the Rio Branco. Sipping hot black tea and writing some journal notes, I was enjoying the cool air when someone shouted my name. Just 300 yards in front of us was the head of an animal in the water, going from right to left. Carlinho said '*cachoho*', but it was clearly no dog. Manuel suggested capybara, but I already knew what it was. It was an adult male jaguar and it was swimming effortlessly, despite the distance and the speed of current. As usual I grabbed my camera. We circled the jaguar and at one point I could have leaned over the bow and touched it. It snarled at the boat. I told Manuel to drop back so that we could watch it emerge on the other side. Gordon eventually appeared just in time to see the strikingly marked cat clamber into the low branches on an island of flooded trees.

A few hours later, with the sun directly overhead, we crossed the equator without ceremony and came to the place where we had left our scaffolding. The jabiru storks were nowhere to be seen and closer inspection under the tree revealed no clues. They had either left the area or been killed. All hands set to carrying and loading the scaffolding on board.

Late that afternoon, our task completed, I lay in my hammock, occasionally looking up to see if anything was happening in the distant but gigantic samauma tree.

'Good heavens, Gordon, it's snowing – look,' I shouted as millions of white flakes floated through the air from just above the river surface to at least 90 feet high. The flakes turned out to be mariposa moths and we realised we were witnessing a synchronised hatching. I wanted to try to capture the remarkable scene on a photo and rattled off half a dozen shots. While looking through the camera behind the 'snowstorm' I saw a jabiru stork flap into the upper boughs of the tree. It had a stick in its beak. I couldn't believe it, the birds were building a new nest just as we'd finished loading all our scaffolding back on the boat. Making wildlife films can be too frustrating at times.

The diesel engine being fired up at 5.30 in the morning brought a fitful night's sleep to an abrupt end. Almost every square inch of the deck and passageways of the *Natureza* were covered with the confetti-like bodies of white mariposa moths. We were now heading for the Xixuau where we would spend the next three weeks, returning to Manaus – hopefully – with the film in the can. Sitting on the prow of the boat I suddenly realised how I could get some decent shots of giant otters underwater. We could ask Marc to bring his pet male otter Frankie – it was the perfect solution. Returning to Manaus would only put two days on the journey and I knew that Marc would be delighted to help. And so it was that three days later we were all heading up the Rio Jauaperi with a seven-foot long giant otter called Frankie flopping about our feet on the lower deck. He was such a calm and gentle natured otter and loved to be stroked and petted.

Our arrival at the Xixuau unfortunately coincided with the Amazon Association taking a tourist group to the place. Twenty Americans and Canadians filled the Indian-style molucca hut (a huge round open-sided thatched house) that the Association had built. We decided to keep well clear of them and moor out on the main river channel. Relations had become slightly strained.

By six o'clock the following morning Marc, myself and Frankie were in our metal canoe motoring up the creek. Frankie was in the bow, staring longingly at the water, just waiting for us to stop so that he could dive in. I hoped to be early enough to avoid any disturbance from the Amazon Association visitors. We unloaded the filming and diving equipment while Frankie had the time of his life exploring the place. When all the gear was ready I sank down to the white sandy bottom of the creek to test the camera. It was only 12 feet deep and extremely clear. Frankie gave me a shock when he suddenly appeared in front of me and thrust his hard hairy nose into my diving mask. I turned on the camera as he twisted and turned. Swimming about me with such suppleness, he looked supreme under the water. I followed him with the camera for a few feet but then he disappeared from my sight.

As I turned back to the middle of the river, I looked through the camera viewfinder to see three furry brown faces staring at me. I couldn't believe it – they were wild giant otters! They were keeping their distance at about 30 feet but all three were standing upright in mid-water – just staring at me. Then, as suddenly as they had appeared, they were gone. I surfaced and Marc could hardly get the words out he was so excited. He had seen it all happen from the surface. Before the three had come to take a look at me, Frankie had met up with them on the land and one of them had been especially friendly to him. What a fantastic experience and what great images.

We frolicked a while in the water with Frankie before setting back. The water was deliciously cold and the sun extremely hot. We watched several times as Frankie quietly left the creek and went into the forest edge behind the small sandbank, where our boat was beached, to meet one of the wild otters. On one occasion they touched noses as they investigated each other. It would have been a wonderfully safe place to have eventually rehabilitated Frankie except, as Marc pointed out, he would have been used as a tourist attraction by the Association when they were visiting and, when not, no one would look after him.

Over the following week we filmed the *castanha,* better known as the Brazil nut tree. The tree bears large wooden husks each about the size of a small melon, each with 20 or more seeds, and there is a tiny hole at one end of the husk. Local people harvest the crop each year as an important supplement to their meagre diets – they are rich in nutritious oils – and to trade on the rivers. They use a sharp machete to open the husk and expertly peel the seeds with the same weapon.

Like all seeds, the Brazil nut needs to be dispersed and there is really only one animal that can do this successfully, the agouti. It is a large, long-haired rodent that looks like a cross between a squirrel and a hare, and has specially adapted teeth for gnawing through the hard husk. Gordon and I built a tower into a tree that was fruiting behind Carlitos's house and set a hide up near the base of the tree where there were several unopened fruits lying on the ground. Within 20 minutes I saw an agouti begin work. First it peeled the softer outer bark of the pod and then rotated it until the small hole was on top. It then started to pare the wood away as expertly as a cabinet maker with a chisel. Any sudden sound like a falling branch would bring it to an abrupt halt, head up, sniffing the air for danger. The agouti is a favourite food of forest people.

It may eat the first nut or seed it extracts but all the others are carried one by one, sometimes up to 100 yards away, and are cached underground at the perfect depth for germination. Using its great sense of smell the agouti will return and recover some of the seeds later in the year. The remainder eventually germinate and produce the next generation of trees.

We moved from filming nuts to vines the following week. On one of our early forest explorations Carlitos had stopped at a liana and, cutting it with his machete, had drunk the clear liquid that poured out in surprising quantity. I thought it would make an interesting item in the film. We arranged to meet Carlitos and his son for them to take us along one of the trails where he knew there was a vine. Carlitos explained that it was important to make the top cut first to stop the water racing back up the vine. After the top cut he severed a three-foot-long piece of the thick liana and held it up over Francisco's mouth while he drank the water that gushed out. I tasted it too. Although the water had a definite 'vegetable' taste it would be a welcome source of drinking water if you were far from a creek. That afternoon's filming went well and I asked Carlitos if he would mind coming out again the following afternoon. He agreed.

The next morning the Association's tour group departed for Manaus and someone gave Carlitos a parting gift of a bottle of *cachaça.* At 2.30, Gordon and I paddled over to his house and found him well drunk. He staggered along the trails into the forest, falling over more than once, his young son Francisco laughing at his state. The light was perfect and I was determined to get the close up of him cutting through the vine. I did wonder, though, if he were in a fit state to wield a machete. The shot made it into the film and it never ceases to make me laugh when I see his face and remember the moment. He faced the camera unable to focus his eyes, and swiped a few times at the vine. When the machete finally cut through the liana, Carlitos wasn't expecting it – in fact I think he had even forgotten why he was there by the third blow. The result of the blade suddenly cutting through took him by surprise and he almost fell over again.

By the end of November we were for once looking forward to returning to Manaus. I had a new home to move into and an expedition with Marc to the Sataré Maues Indians, where we would research the marmoset film, to prepare. The house at Igarapé Jacaré was even better than I expected. We arrived to find the men putting the finishing touches to the three small internal rooms that were to serve as two bedrooms and an equipment store. The open-plan living and hammock sleeping area was palatial and the view from it across the creek to the forest wonderful. We humped the first sticks of furniture we had brought with us from Manaus up the sandbank to the front door. They were garden furniture table and chairs. After Gordon jokingly offered to carry me across the threshold, we wasted no time in setting them out, slinging our hammocks, opening two bottles of beer and, clinking bottles, toasting my future here.

'To your success, too,' I saluted Gordon. A part of me felt sad because we both knew that it was only me, not the two of us who would be living here, but Gordon was looking forward to his new challenges with the right spirit, and to visiting me here in the future.

I gazed around the inside and tried to imagine that future – how many years? – how many films? – and what kind of person would my next assistant be? In my mind I pictured Emma swinging in a hammock, candlelit evenings with scripts and books, listening to *The Sound of Music*, *South Pacific* and other nostalgic songs on my CD player. The atmosphere in the house felt right.

Marc had only returned a fortnight before from the Maues area where he had spent a week researching our marmoset project. With only one week until our planned joint marmoset expedition we had just a few days to boat into the house the other bits and pieces such as 40 gallons of wood preservative (to keep the termites at bay), water tanks, cement, water pump, fuel tanks and – truly a prize find – a propane gas refrigerator. There were also several tons of scaffolding, pulleys and ropes that would take four or five days to transport. This we decided to do with a larger boat along the main river to the mouth of the Jacaré creek and then ferry it bundle by bundle to the house in our metal canoes.

There were no windows and all four sides of the house were open as though it were one enormous veranda. One nice touch was that because many of Marc's orphaned monkeys would be released here, we encased the entire house in wire mesh so that it would be the humans who would be living in a cage while the animals roamed free. Mind you, there was little choice because the alternative would have been a 'war of the primates' every minute of the day – precious few creatures are as destructive as 'tame' monkeys.

Another task was to look for two *caseiros* – handymen and resident caretakers – for the camp. Back in Manaus Marc and Betty had recently had the doors and windows of their house renewed and the carpenter who did the job said he was interested. He was 34, told us he had no family, read the Bible, and could climb trees. Judging by his workmanship he seemed perfect. We agreed to employ Manoel full-time. As soon as we told him we would be arranging a second worker he told us that his nephew would be ideal. Less than two minutes ago he had had no family, and I wondered how long it would be before a wife and ten children surfaced. His nephew Gilson was strange, to say the least. I stuck my hand out to shake his but he either couldn't see it or ignored it. I pulled a funny face at Manoel in a 'what's up with him' gesture but Manoel just laughed and said Gilson would work well under his direction. The two of them would build a second smaller house a little further along the creek from mine. I had a feeling that Gilson wouldn't last very long.

Chapter Sixteen

The Sataré and a Discovery

Everyone called Valdemar 'Gordo' because he was fat (*gordo* being the Portuguese word for fat). He was our Mr Fix-it for the reconnaissance trip to the Sataré Indians. He has travelled the rivers and creeks in Sataré territory for many years, trading mainly cotton material in exchange for guaraná seeds. Guaraná is a popular drink in Brazil and is made from the crushed seeds of the guaraná plant. When in fruit it looks as if numerous eyeballs are staring out from the tree. It is supposedly an aphrodisiac.

Gordo was short as well as fat, with a podgy face topped with thick black wavy hair. His eyes always looked as though he had a monumental hangover, heavily veined red and yellow. He talked with expansive arm gestures – and he talked a lot. Despite all of this he was easy to warm to and had a great sense of humour. On his chest he had an eight-inch scar which he told me was caused by the Indians when he first met them. He had, in fact, been stabbed and slashed in a bar brawl.

Our boat for the expedition was the *Morreira* and its owner, Fernando, and his wife, Francy, the cook, were coming with us. They had two young children, one three years old and a six-month-old baby. I didn't think it a good idea for them to be on board, but Gordo had agreed to this when he had rented the boat on my behalf because Fernando was cheaper than the rest. Fernando didn't have a licence to sail and so had employed a skipper. Francisco da Silva was 60, looked 70, was painfully thin and bad tempered, too. I didn't think it a good idea to have him on board either, but we needed a legal captain. His first words to me were to ask if I had any *cachaça* with me. With Gordon and me were Marc and his youngest son, Tomas, whom we seconded to operate the sound recorder because Gordon would be using the second camera.

FUNAI is the Government authority set up to look after the interests of all Brazilian indigenous peoples. Even to enter an area where Indians live requires its consent. The previous day Gordo had confirmed by radio that our visit was to be allowed but only if, at the Sataré chief's request, we brought a cow. Where we were going to get one and, more importantly, where we were going to put it on the 30-foot *Morreira*, I didn't want to think about. Gordo assured me that everything was '*tudo bem*'. The round trip by boat to the Rio Andira was getting on for 300 miles, the first and last 60 on the Amazon itself. The memory of our last ill-fated river expedition down the Amazon where we almost drowned was still fresh in my mind and it was with no small trepidation that I stood at the bow rail of the *Morreira* with Marc and Gordon as we sailed across the confluence of the Negro and the Amazon. This boat was a little smaller than the *Natureza* and we seemed so tiny in the middle where these two massive rivers met. The first thing I had done was to get the owner Fernando to tell Mr Grumpy at the wheel that under no circumstances were we to sail after dark.

I went below to use the toilet and discovered one of the reasons the *Morreira* was probably cheaper than other boats. There was no bathroom with shower, only a toilet – and that could only just claim that title. I almost had to get on my knees to get into it.

Our first night aboard was simply awful. We had tied up to the river edge but the wind ripped across the two and a half mile wide Amazon, slamming into our starboard side. The boat rocked constantly, throwing the hammocks from side to side. After crashing into the bulkhead and then Gordon several times, we both gave up and went out onto the rear deck. Under a brilliant, starry sky we sipped vodka and Sprite and talked until almost three o'clock. I woke Fernando, who somehow managed to be sleeping well, and asked him if we could find a *paraná* (channel) where we would be more sheltered. Francisco was in a frightful mood and stamped up to the helm while the engine was fired up. Gordo told us that everyone called Francisco 'Onça' because he had a temper like a jaguar.

Half an hour later we were snug inside a *paraná* and the next thing I knew it was almost eight o'clock in the morning. Gordon was organising coffee while I braved the ship's toilet. Sitting there, still half asleep, the door swung open to reveal me in profile as Francy, boiling water on the stove, stared and then laughed. What a start to the day.

Early afternoon we entered the Rio Uraria which joined courses with the Abacaxi (Pineapple) river 40 miles south. It was immediately noticeable that the few houses we passed were very different from the usual Caboclo construction. These small dwellings were made entirely from palm thatch panels and put together in a most attractive way.

Gordo appeared waving his arms saying that we were going to stop shortly and try to negotiate for a cow. The hut we pulled up to was almost in the water but 150 feet along the bank under the shade of a copse of palm trees were about 20 zebu cows. The only person in sight was a woman fishing from a canoe on the other side of the river. She was about 60, with a surprisingly hairy face and a shock of jet black hair that was so long it almost touched the water. Holding her wooden paddle, which she handled expertly, she reminded me of a Halloween witch. She told Gordo that the owner of the cows was in the forest but should be returning shortly. An hour later the crew and the local man were struggling with the roped animal trying to get it on board.

That night, with a cow tethered to a post in the dining area of the lower deck, Gordo started to explain what this was all about. The Sataré Indians stick fiercely to their old ways and rituals, although most of their group have been contacted by missionaries. Once a year all the villages in an area get together for a dance, but this is a dance with a difference. The head man of the village we were going to knew we were coming from a big town and had decided we could save them hunting tapirs and monkeys for a week to provide enough meat for the feast that always follows the dance if we brought a cow.

The following morning we were sailing along the Abacaxi when Gordo pointed to a small group of palm huts at the next bend. It was the Sataré village. The five huts were on a small bluff and a crowd of Satarés were watching our arrival with fascination. As Fernando threw a rope to one man a great cheer chorused from the crowd – they had spotted the cow. These small lean people with dark brown skins and jet black hair whooped with joy. They smiled openly and welcomed us as friends. A young Sataré man climbed aboard and spoke to Gordo, which he translated as '*elles vão dançar amanhã*' – they are going to dance tomorrow. While Gordon and Tomas prepared camera and sound equipment for the following day, Marc and I were taken up to meet Chief Diko.

The crowd had increased to about 70 people. They parted, chattering excitedly and laughed as we clambered up the mud and grass slope towards the huts. As though our leaving the boat were a signal, a group of young men moved in to unload the cow – I didn't want to watch. We were taken into the largest of the palm huts and the chief looked as though he was sitting on a wooden bench. It was dark inside but my eyes adjusted quickly. I soon realised that he wasn't sitting, he was simply tiny. He spoke good Portuguese and Gordo introduced us. Motioning us to sit down he started to talk slowly, saying we were welcome and that we would enjoy tomorrow, especially when his people initiated the young men with the Tocandeiro Ant Dance.

Outside, a group of men and women were singing tunefully and I could tell that the atmosphere was charging up. With daylight fading fast, we shook hands with the chief and returned to the boat. On the slope at the side of the boat was the cow waiting for the inevitable end. A group of ten young children were sitting on the ground close to it. We ate our own meal of beans, rice and chicken on board that night and talked about how we were going to film the following day. After washing from a bucket off the stern of the boat, I went to my hammock at ten o'clock feeling excited and hoping to get a good night's sleep, to wake fighting fit for the hard but interesting day ahead. Sadly, it didn't happen like that. The Sataré's annual festa had started and was in full swing. When they let their hair down, they carry on all day and night – for three days! Decent sleep was impossible with the howling and singing, and the bamboo drums and flutes banged and droned away all night long. At 5.30 am I heard the cow being killed and put my fingers in my ears.

At seven o'clock Chief Diko sent word that the men were about to go off into the forest to collect the tocandeiro ants for the dance ritual. We hoisted our packs onto our backs and set off behind seven young men, their faces painted with black markings. After 15 or 20 minutes they gathered around the base of a tree and two of them began smacking its roots with machetes. Within seconds the biggest, fiercest looking ants I have ever seen began to crawl out of their nest to look. Two-inch long ants with pincers like pliers didn't seem to be things to mess with, but the men prodded away and the ants bit into their sticks with gusto. They were then lifted over the mouth of a three-foot-long hollow bamboo tube and tapped into it. Every so often one man would tap the bamboo pole onto the ground a few times to keep the ants at the bottom. Tocandeiro ants are not only armed with formidable jaws, they are also highly venomous. The common English name for them is the '24-hour ant' simply because if you are stung by one that's how long the excruciating pain lasts.

Two of the young men blew noisily on small bamboo flutes. They needed to collect about 300 of these giant ants and had to raid several nests. We filmed the whole process for over an hour in the sticky humid forest. The Sataré laughed constantly and seemed to be telling each other jokes.

Back at the village, in the central area between the huts, two men with hands and wrists painted black were preparing a bowl of water. One crushed a handful of leaves from a cashew nut tree into the water and the other stirred them with a stick. After a few minutes the chief unplugged the bamboo pole and tipped the furious ants into the bowl. The effect was instantaneous and the black mass of gigantic ants floated motionless, temporarily drugged. A young man approached with a tube-like container woven from fresh, green palm fronds. The two men who had prepared the liquid squatted and began to pick ants out of the bowl. In the woven tube a man then pushed a small wooden spike the size of a pencil through a palm frond, and into the hole that it left, another man pushed a tocandeiro ant. They took 40 minutes to weave all the ants into the palm tube. The ants' abdomens, with their dreadful stings, pointed inwards while their ferocious mandibles remained on the outside.

It seemed everyone in the community looked on, many with painted faces. The only colour used was black, the dye extracted from the genipapo fruit. Some women wore jute cloth dresses and macaw feather headbands. A few young men stood out from the crowd because their hands and wrists were painted black. These were the initiates. The palm tube is actually the inner lining to a glove and that glove was brought forward with much oohing and aahhing from the gathered crowd. The chief held up the palm tube and blew upon it. The ants bristled with anger – they had come back to life with a vengeance. The chief placed the tube inside the larger decorated glove topped with brilliant red and blue macaw feathers.

The light was poor, it was hot and the sky was obliterated by total cloud cover, but for us the scene was electrifying. Marc, Gordon, Tomas and I kept looking at each other, knowing what was coming. Gordo simply kept shaking his head as though they were all mad. It was difficult to believe, but eight young men were going to thrust their hands into that glove and suffer the consequences. That's how you prove yourself as a young Sataré man and guarantee a long life!

To the side of Chief Diko's house was a communal open-sided shelter, its roof made of the same palm panels as the other houses. It was about 50 feet long and was supported by many wooden poles. Running straight through the centre from one end to the other was a thick chest-height bamboo pole. Twenty-five dancers stood along its length and the glove was placed on an upright pole in the centre. A young man stepped forward from the line and held out his right arm, resting it on the bamboo pole. Another man, who appeared to be the master of ceremonies, lifted the glove off its pole and placed it on the initiate's hand. The initiate's face grimaced instantly and his head drooped, but he never made a sound. The two dancers on either side of him gripped his forearms. All at once the entire line-up was swaying backwards and forwards and everyone began to chant. The young Indian maintained an expression of agony, his feet slamming the earth floor in time to the beat of the music. This he kept up for five minutes before lowering his hand onto the bamboo banister and having the glove removed. Then the next young man moved forward to receive the glove.

The whole business was mesmerising to watch and, of course, film. I wondered what on earth the first missionaries had thought of all this. One thing Gordon and I agreed upon was that, no matter what, we wouldn't try this one out of respect for culture. Several of the young men who had been through the rite eventually moved out of the dance group and, moaning loudly, wandered through the crowd holding their painfully swollen hands up for all to see. They were given a stick of tree bark to smoke – its effects hallucinogenic, we were told. A vat of fermented manioc hooch was also being passed around and all were eagerly glugging it down.

Many of the visiting Sataré enjoying the ritual were from a Baptist and Catholic mission an hour further down the river at a place called Porto Alegre, 'Happy Port' – as it was named by the missionaries. The chief asked Gordo if we would take them back to their village later and, of course, he couldn't refuse. We stayed among them until late afternoon. When the last young man had done his bit, the Sataré were still dancing, eating cow and singing, while we all dived into the river to wash off three days' dirt.

'Wouldn't a nice clean bathroom be the ticket now?' I said to Gordon as we lathered our hair in the muddy water.

As we climbed back on the boat it was besieged by 30 or more of the Sataré who were now ready to be taken back to Porto Alegre. They swarmed all over the top deck, making the small boat roll alarmingly. On the way down river Gordon was deep in conversation with two young Sataré girls. After we had dropped the gang off on a beach in front of a huge cement church and were making our way back to Chief Diko's, he told me that the two girls had invited him to sleep with them at the mission station. He, being a true professional and sensible chap, had decided it more prudent to record the night-time singing back at the ritual.

The main reason we had come to the area was to find out about marmosets and what species lived there, and the '*Dança do Tocandeiro*' had been a bonus. We knew that many Sataré Indians kept the miniature monkeys as pets, but we hadn't seen a single one at Chief Diko's small community. The following morning, Marc went to speak to the chief and his villagers to find out what they knew. Three species live in Sataré territory but are isolated from each other by two main river systems – to the west the Abacaxi and Uraria, which eventually join, and to the east the Maues river. I was especially interested in the golden white tassel-eared marmoset. It is nearly all white and is a striking monkey that looks like a gremlin. Marc came back with good news. One family who lived in the community had a brother, Josico, who had built his own place about 12 miles north. He had several children and usually kept marmosets as pets. Only a week ago, the chief told him, he had had three in his own house but had traded them to another Indian.

On 4 December we said goodbye to Chief Diko and his people and set sail for Josico's place. After three hours we turned into a narrow channel. The trees at the river-edge brushed against the side of the boat in places and occasionally large iguanas that were basking in the upper branches leapt from the trees, free-falling to the water below. We saw three small thatched huts near the water's edge and two children playing in the water. A large tree whose base was underwater provided a convenient mooring post and Fernando pushed out the gang plank

Josico was coming towards the boat followed by two young girls and their mother. Two little boys – twins called Mezaki and Zatraki, we later found out – were standing under a thatched shelter where wisps of white smoke floated up from a fire and seeped through the thatch. Small broadleafed cashew trees provided some shade in the clearing between the huts.

Josico could speak Portuguese and as Marc started to explain our mission Mezaki came over to my side and tugged at my arm. His brother watched him do this and then came over himself. He pulled over my wrist, closely scrutinising my watch. Josico got up and, motioning us to sit on logs at the edge of the shelter, went over to the fire and fetched a large metal pot. 'I hope it isn't açaí juice,' I said to Marc. Zatraki tried to mimic my strange-sounding words and the two boys burst out laughing. The drink is made from the dark purple fruit of the açaí palm and I hate it. A minute later, though, I wished it was açaí.

Josico put the pot on the earth between us. It contained thousands of sauva ants. Better known as 'leaf cutter ants', these large-headed creatures are common around *roças* (farms) and the Indians regard them as a delicacy. Marc dived in and chewed away.

'You'll have to eat them, Nick. Actually I like them,' he pronounced. Mezaki and Zatraki didn't waste any time either – they clearly loved the ants. The pot looked horrible but I picked it up and sniffed the contents. Surprisingly I found the smell agreeable. It was a heavy lemon scent.

'OK, how do you eat them?' I asked Marc and he explained that you simply bite the large head – with ferocious looking jaws – and suck out the liquid. So I bit and sucked and the taste was dreadful. Seconds later my tongue and lips felt numbed. The liquid is formic acid, hence the lemony scent. When the ants attack something and cut with their huge jaws, they spray the formic acid into the wound.

At first Josico confirmed that he knew three different species of the little monkeys and that his daughter and older son, who would be returning later that day, had two infant pets. As Marc questioned him further and asked him to show on our map where he hunted and sometimes saw the monkeys, things became confusing for us. We had pictures of the three known species, but Josico maintained that he only recognised two of them and that the third one he often saw was not in our pictures. The main difference, he said, was that it had bare ears and not big fluffy ones like the others. It became clear that there was another kind of marmoset here and it slowly dawned on us that we might be on the verge of seeing a new species. Not new to the Sataré, of course, they have lived in the area for thousands of years, but new to the scientific world at large. We were eager to see this monkey and to our great excitement Josico told us that his daughter had one of them.

We sat and chatted through to the afternoon until we saw a canoe approaching. Even from the distance we could see the white marmoset tethered to a line and darting about on the edge of the canoe. What we couldn't see at that moment was the other tiny creature in the thick black hair of the young girl sitting between the two young men. As they left their canoe and walked up to the shelter Marc looked amazed; he could tell immediately that this was indeed a new species. I couldn't believe our luck. Josico took us across the river and into the forest where he usually saw the group. Having seen the white marmoset close up, and with this revelation of a new species, I had no doubt that a film should be made and that Survival would be as excited as we were about it.

We stayed at Josica's overnight on the boat and early the following morning waved the family goodbye for the first leg of our return journey to Manaus. It was a bad start to the day for me because my last tube of British Crest toothpaste had run out and I had to use some of the awful Brazilian Kolynos which left a horrible taste in my mouth. It probably seems pathetic to most people but those little things – familiar, comfortable things from home – mean a lot to me out in the wilds.

Gordo and Marc wanted to stop at Santa Rosa on the Uraria so that they could say hello to an old friend they had made on a previous research trip. Santa Rosa was one hut, and a small one at that. The owner was a tiny, hermit-like character who liked to be called Sr Ratinho (Mr Little Rat). His real name was José Motta, and he had lived alone here for 15 years. As the boat was manoeuvring between two trees in front of this isolated place, the diminutive Sr Ratinho appeared from under a small thatched shelter. His face was screwed up as he looked through two thick frosted bits of glass that used to be a pair of spectacles. He didn't move an inch, just stayed where he was until Gordo was a few paces from him and shouted out loud 'Sr Ratinho'. The old man darted forward, grasped Gordo's arm and shook it almost violently. He spoke ever so slowly with a croaky voice.

Marc stepped forward and greeted him and Ratinho was so happy he literally jumped up and down on the spot. Marc handed him some photos that he had taken of him during their first encounter. Sr Ratinho dropped onto his knees and spread the snaps on the earth. With all of us looking on in wonder at this character, Sr Ratinho kept saying 'ooohpa, ooohpa' in ecstasy. He was obviously lonely and completely fazed by so many people dropping in. Gordo introduced me to him and he gripped my right hand tightly – and wouldn't let go. He also gripped my right forearm with his left hand just to make sure I couldn't get away.

Like this he half walked and half pulled me over to the shelter where I had to duck to get inside. Sr Ratinho only came up to my chest. '*Água ... para café*,' he shouted slowly to everyone and, still holding on to me, led me towards the open fire. Here at last he let my hand loose. He got the fire going and put a pot of water on to boil when Marc came over to us and told him that he had another present for him. Marc had realised that Sr Ratinho's eyesight was extremely poor and that his ancient glasses were doing him no favours. He had bought a pair of new ones in Manaus, just guessing at what would do, and he handed them to the old man. He put them on with great ceremony. A whole new world seemed to open up for him and he 'oohed' and 'ahhed' for five minutes while he walked around his familiar home picking up objects, scrutinising them an inch or two from his face, and then putting them down. He then came up to each of us and peered into our faces, able to see better than he had done for many years.

While an argument raged between Fernando and the bad-tempered captain about us being short of fuel, Gordo explained that we were about to leave. I felt sorry for him because he had enjoyed the company so much. As he shook my hand, gripping my forearm again, I asked him how old he was. '*Deixe me pensar ... Eu não sei ... mais de setenta*,' let me think ... I don't know ... more than 70.

Our sail home was a struggle. The Amazon was in full flow heading towards the Atlantic and our boat only just made headway against the current. Worries about fuel had everyone on edge because you can't just pull into a gas station to fill up out there. In-crew squabbles added to the poor atmosphere. Sourpuss at the wheel was constantly fighting with Francy because she wouldn't wait on him hand and foot. Tomas had fever and felt extremely unwell.

Later one afternoon I went to see how Tomas was feeling and found him squeezing an unsightly lump at the top of his left thigh. A great deal of blood and disgusting green pus came out and then – a two-inch-long white worm! Within 24 hours the fever had gone and he felt tip-top.

During the last hour of daylight Gordon and I sat on the bow ladder just soaking up the river and forest scenery, counting birds of prey. One, a small black hawk, rose into the air from the undergrowth with a brilliant green parrot snake dangling from its talons. Francy then appeared and the fight with Francisco began again. Gordon started to sing 'If you're happy and you know it clap your hands,' and we laughed looking at Francisco who had a face like thunder. 'Miserable old bugger,' Gordon chimed.

Our last breakfast was slammed down in front of us. Francisco's rantings had finally taken effect. Eggs soaked in oil, stale bread, wedges of greasy cheese and bricks of processed ham were washed down with sugar pretending to be coffee. By now, I hated coffee. The boat suddenly veered, thankfully spilling mine. Fernando legged it up the ladder to find out what was going on. Being the inquisitive types, and sensing trouble, Gordon and I followed. Old jaguar chops was pulling one of Plinio's old tricks. We were perilously close to the stern of another, much larger, boat. I exploded – verbally. It is extremely difficult to master anger in a foreign language, but I tried. I could say 'Sonofabitch' perfectly – the sentiment is much worse in Portuguese than English – so I said it. Francisco just stared, face set in defiance, dead ahead. Only when Fernando, for the first time, spoke to him in an obviously angry manner did Francisco swing the boat off to one side and safety.

'How the hell do you say half-wit in Portuguese?' I asked Gordon.

'Francisco,' he replied.

Chapter Seventeen

New Year – New Home – New Films

Gordon and I both returned to Britain for Christmas. I had a brief two weeks to enjoy family, friends and the Christmas spirit – well, only the traditional spirit for me. On 24 December, 1993, I developed dreadful abdominal pains and had to start a course of amoeba medicine, which meant I couldn't take a drop of alcohol. At least I had Emma with me and, despite the waves of stomach pain, when she would hug and try to nurse me like a miniature Florence Nightingale, we had a terrific time. I managed to get a telephone call through to Manaus and we both wished Antonieta, Marc and Betty a happy Christmas in Portuguese. Everything was going well at the new camp and Manoel and Gilson were staying there over the holiday period.

Before we knew it, Gordon and I were hugging each other and saying goodbye to our five and a half years together. We had shared a wonderful experience and would never forget it. But new challenges faced us both and it was important to look forward, not back. I promised to keep in touch and do anything to help him find his feet. (We have kept in close contact ever since and he is making his own way in the freelance film world.)

On 8 January, 1994, Pete, Ama, Yetta, Pipoca, Lucy, Tarzan and Jane moved in to the Jacaré with me. Pete, Ama and Yetta were spider monkeys, the rest woolly monkeys. Pete was an infant, Ama and Yetta adult females. They were all special, but Yetta was handicapped and ignored by the rest. She, however, made sure that we couldn't ignore her and many visitors to my camp fell in love with her in particular. She had a sad history. On two occasions her head had been split open by a maniac with a machete and she had been left to die. That she survived was a miracle. The injuries had left her with brain damage and mild epilepsy, but for most of the time she was extremely happy to clamp herself around any part of your body. She particularly liked legs, and left you hobbling around with a big black furry spider-monkey moon-boot.

Apart from the monkeys, there was Gordi the baby tapir, Bambi, Flower and Biscuit the deer, and a flock of parrots, macaws and toucans that were better at waking us up than an atomic alarm clock. All these animals came from Marc and Betty's project where they took in orphaned and confiscated animals from IBAMA and tried to give them as good a life as possible.

The Jacaré couldn't have come together at a better time from Marc and Betty's point of view. They had had major problems with the people who owned the land where they ran their project and had literally to run for their lives one afternoon when 'pistoleiros' appeared and put a gun to Marc's head. Having fallen out with the owners, the Roosmalens were forced to leave and planned to return to Holland. What to do with their animals was a real concern as they obviously didn't want them to end up caged in zoos. Then came the possibility of us working together on some films, and the forest camp provided a perfect safe haven for some of their creatures.

There was never a dull day at the camp. Within my first week the spider monkeys had learnt how to break into the house: we had to place stronger wire grills over all the openings. I spent the first two months building the new filming towers into the forest canopy, with Pete, Ama and Yetta trying to help every inch of the way. Pete would be perched on my shoulders, or dropping hammers 150 feet to the floor while Yetta clung limpet-like to one of my legs. Still, it turned out to be a great advantage having them about because they found all the fruiting trees and alerted us to other wild groups travelling through our area, and to any possible dangers. Both spider and woolly monkeys create a tremendous din when they are upset. The spider monkeys especially start duetting to each other and the alarm barks resound through the forest as they grow in intensity. It's quite wonderful to hear although nerve-wracking if you don't know what has set them off. It could be anything from a harpy eagle to something as harmless as a cat.

By the end of March 1994 'grunter' Gilson was, as predicted, long gone and we had a new worker. Claudio was 28, gangly, cross-eyed and spat when he talked. The rota of time-off for the men revolved around 11 days working on-site and three days off in Manaus, and for most of the time this scheme worked. We had cut a trail system for Marc's tree research and had built three towers, two of them poking above the canopy at 150 feet, giving a fantastic panorama of the vastness where we lived.

The most impressive construction for me, though, was a sit-down loo – the years of pit latrines could now be history! Manoel built a 20-foot wooden tower and on top of it we placed a 500-gallon water tank. It was plumbed directly to the house and to the base of the tower which we panelled in and, there, fitted a shower head and a toilet. The supply for the tower tank came from the creek and we pumped up fresh water every two days with a small portable petrol-driven pump. The bathroom also needed to be wired in after the monkeys discovered what fun it was to roll out miles of loo paper into the surrounding forest, eat bars of soap and swing on plastic water pipes until they broke. Manoel was an excellent worker, but rather than tell me he didn't know anything about plumbing he tried to muddle through. For some reason he connected the top of the home-made septic tank directly to the shower drain in the middle of the small cement floor. As you can imagine, the result, in just a few days, was simply awful! We had to dig up the floor and rework the whole system, but eventually it functioned well and was a blessed refuge at times.

The house was also a shelter from frequent tropical storms for us and our primate friends. The roof was built with a long over-hang to help prevent driving rain coming in and the monkeys would cluster on the wire grilling underneath it peering in at us and, more often than not, begging luxury food items like chocolates or our home-made yoghurt. We made tarpaulin blinds that could be dropped in the most severe weather. On dark, stormy nights, with the sky constantly lit by lightning flashes and the sounds of forest giants crashing to earth nearby, we would sit snug inside where candles flickered. It reminded me of winters on my island home in Scotland when the Atlantic weather system threw its worst at us.

There weren't many nuisances, apart from certain insects. Thankfully, mosquitoes rarely bothered us, but other smaller biting flies such as mole crickets with fearful bites, and wasps, were irritants. One morning in early April I took a T-shirt off the wall hanger and pulled it on over my head. As I combed my hair in front of the mirror I felt a tickle on my left shoulder. I scratched it but it returned and as I looked under my shirt sleeve there was a small jet-black scorpion, its tail raised ready to sting. I nearly jumped out of my skin and stupidly brushed the critter off. Luckily it did not sting me. But only a few weeks later, Marc was stung by one when he came to the camp. It left him for almost 24 hours with a paralysed arm and in a great deal of pain. From that day on, I always shook my shirts, pants, socks and boots violently to expel such visitors.

With a permanent forest home I could at last welcome human visitors and offer them a degree of comfort. First to visit was Bonnie, an old friend who was a long-haul purser with British Airways. Bonnie is one of the most seasoned travellers you could meet, but the Amazon rainforest was new terrain for this Lancashire lass. It turned out that she is one of those unlucky people who, despite being covered from head to toe in body armour and insect repellent, simply oozes whatever the pheromones are that attract flying biting bugs. However, she never complained, and her 'Red-Cross' hamper bursting with recent English newspapers, Travel Boggle (a great game), Earl Grey tea, chocolate Hob Nob biscuits, and more, were received extremely well.

'I expect it's back to living out of a suitcase,' I said to her at the airport when she was leaving. 'Not likely darling – I'm going to soak myself in a health farm for a week to get rid of all the bite swellings, but I'll be back one day,' she promised.

My business partner, Nick Peake (together we set up an independent film production company in 1988), and his partner Mary Rose Kane (Kanie), were next to visit. Nick and Kanie both worked in Granada Television and so were intrigued to see how wildlife films were made – a slight difference from their social rights programmes and popular game shows!

As Bonnie is one of Kanie's oldest friends, naturally she had given them the low-down about the place before they left England for my Amazon camp. Among other things, we all share the same passion for Boggle and music. We Boggled morning and night with 'Master Boggle', the ultimate game, throwing in a few sing-songs as well. Nick is an accomplished guitarist – as well as an infuriatingly good Boggler – and we had tremendous fun entertaining the monkeys who, hanging onto the window grills, chuckled and chattered along with us.

Antonieta, who at this time was working for an air cargo company at Manaus airport, took her holidays and came to stay at the camp, too. Within two days the house was transformed. Tablecloths appeared and sprigs of wild flowers adorned my bookshelves. Our diet leapt to heights it had never known and Kanie even introduced 'home-made' pizzas which our jaded palates have enjoyed ever since. On 5 November we built a miniature bonfire by the creekside and, armed with a few tins of beer, crooned in harmony to celebrate Guy Fawkes night under a startlingly clear Milky Way, while Nick expertly strummed our everlasting requests. The surrounding forest was literally alive with the Sound of Music that night – to Nick's continuing embarrassment it is one of Kanie's and my favourite scores.

Kanie and Yetta hit it off from day one and during those two weeks it was a common sight to see my partner's partner lying at the edge of the crystal-clear creek reading a book with Yetta fastened across her back. (As long-standing close friends, Nick and I have developed an in-joke where we call each other 'Partner'.)

Getting Nick up the tower was an ordeal for both of us. There are no ladders, only small metal collars on each corner post. To climb the tower you simply scale the corner posts using the collars as foot steps – and, of course, clinging on for dear life. I have climbed towers hundreds of times and rarely give it a second thought. Only when people who are not used to it have a go, do I realise how blasé I have become. The majority never make it to the top and our golden rule is that if someone isn't happy, then we abort the climb. All those who want to climb must wear our mountaineering harness which has a rope attached to it that passes through a bracket and pulley contraption at the top of the tower. Should anyone slip, someone at the base then turns the rope around one of the base poles for a brake, and takes the weight of the person climbing.

Nick climbed and I kept pace with him on the opposite corner post just to keep him company and give him encouragement. With gritted teeth he told me to 'shut up' as I pointed out interesting things to take his mind off the enormous drop below him.

'Partner, isn't it amazing how the branches below your feet give you a false sense of security up here?'

'For God's sake, shut up Partner and let me get on with it!' he said at 50 feet.

He slowly but determinedly ascended through the middle and upper forest layers to the platform on the top where our monkeys were almost yawning with the boredom, waiting for us to arrive. Those who make it to the top soon forget the climb when they take in the view – it is simply stunning. As Nick crawled over the edge of the metal decking, Pete jumped into his lap and they cuddled while Nick regained his composure.

After my friends had left it was back to serious work: three months of sitting on the tops of the towers trying to piece together the private life of spider monkeys (the first film Survival commissioned me to make from my new base). These delightful furry black and spindly creatures play a vital role in the regeneration of the Amazon rainforest. They eat mainly fruits but, unlike most other monkeys, swallow them whole and so ingest the seeds with the flesh. This does not happen by accident because the trees are smarter than we think. To attract the spider monkey to them, once a year they produce a fleshy fruit that is almost impossible to separate from the seed and so the spider monkey has adapted to swallow – not chew – the whole thing. The monkeys travel great distances during the day and later, miles from the tree that produced the fruit, they get rid of them from the other end. The seeds are dispersed – in a free packet of fertiliser. The spider monkey's digestive juices soften the seeds' casing and so help them to germinate. As Marc taught me, this is only one example of the way trees and animals have evolved to help each other.

Left to its own processes of selection forest life would continue to flourish. But, as elsewhere, we have interfered with the cycle. Many of the larger trees that the world regards as valuable sources of timber are the ones being cut down at an alarming rate and are the very trees that produce the fruits that spider monkeys alone eat to survive. One thing we humans seem incapable of grasping is that if the trees disappear then so will the animals that depend on them. Conversely, of course, if the animals are hunted the trees will suffer the same fate, only we shall not see that sad end because a tree's life span may be more than 150 years. Seeing what is going on at first hand fills me with a sense of helplessness.

I had purposely chosen two of these fruit-bearing trees in which to build two of my towers, knowing that they would be irresistible to the monkeys. I would walk into the dark and damp forest at dawn with my flashlight, with Claudio helping to carry the camera gear. I like the mood of the forest at dawn better than any other time of the day. It is always one of expectation as insects, mammals and birds begin to wake up. The metal towers are invariably wet and climbing them is hairy, to say the least. Slippery poles and rubber-soled boots don't mix and it is a slow process inching up 150 feet.

One morning when we arrived at the base of a tower I was inexplicably filled with a sense of fear. I craned my neck to look up at the top level and it seemed higher than ever before. I touched the cold wet metal pole and drops of condensation trickled down it to the floor. For the first time I decided to fasten the equipment hoist rope around my waist and chest as a safety harness, explaining to Claudio how to make a brake with the rope. At 50 feet my boots slid off a metal collar for the first time and my arms could not hold my weight. I fell backwards away from the tower. It happened so quickly that my mind didn't register the potential horror until later. The safety rope cut painfully into my groin and waist but Claudio had thankfully done as I had asked and I only dropped three feet. The coils around the bottom pole on the tower bit into each other and I swung back into the tower and held on tight. My legs trembled for the next 100 feet and when I reached the top it took several minutes for my pulse to recover. I looked over the edge at Claudio and he shouted to me '*Muita dor nê?*' Very painful no? That it certainly was.

I never became bored on the towers, even if I was there until sunset – another glorious time to be alone in the canopy. I had a routine of watching through the observation slits where I could also write up my notes or snack on lunches of boiled rice and dried fish. The worst part of the day was always between eleven in the morning and three in the afternoon when the heat was at its fiercest and most creatures were resting in the shade below the branches. At other times there was always a great variety of birds, insects or monkeys coming by, either to feed on the fruit or simply calling as they passed. Stunningly coloured scarlet or blue and yellow macaws squawked loudly and sometimes settled in a nearby tree. Screaming pihas, trogons, spangled cotingas, woodpeckers, fan-headed parrots, paradise jacamars and, when I was really lucky, an enormous harpy or crested hawk eagle would visit. At dusk Claudio or Manoel would come and lower the gear to the ground and then I would head back to the camp for a refreshing dip in the creek, followed by dinner and my hammock.

At the beginning of March 1995, I needed to return to Britain and edit the film which was now called *Web of the Spider Monkey*. Twelve weeks in civilisation was a welcome thought after the months of isolation and I wanted to arrange for Emma to come out to the camp in her school summer holidays. It was also time for me to take on a new assistant to prepare for my next project, on marmosets.

For six years Stephen Terry had been writing to me for a job. He is a zoology graduate from Cambuslang, Glasgow, and had been working for two years as a zoo keeper in Singapore. He was 28, thin, had dark curly hair, a good sense of humour, liked beer and turned up on time! He also seemed suited to the lonely lifestyle. I had, in fact, already met Stephen briefly and decided then to employ him. Stephen arrived two weeks before I left Brazil. He came through immigration at Manaus airport smartly dressed in a jacket, waistcoat and tie and – to my surprise – earrings!

At the same time Antonieta had said that she was interested in working on the project and we agreed to give it a try. I was concerned by the prospect of her leaving a full-time career, but the air cargo company she worked for promised to keep her position open should she want to return within the year. There was also the inevitable worry as to how our relationship would manage the transition from 'going out together' to living under the same roof. As it turned out I needn't have worried.

Once in the UK, seven weeks of editing the spider monkey film with Howard Marshall at Survival passed quickly. The favourite part of the day for me was 11 o'clock when the sandwich lady arrived. Fresh crab baps, tuna rolls, cheese salads and traditional bags of crisps with little blue bags of salt in them were simple fare that I had missed for so long. With the film 'rough-cut' I went up to Tobermory to be with Emma and to write the script. She was thrilled by the prospect of coming out to the camp, especially after she had been granted two extra weeks of holiday to make the trip in August.

On 10 August, Emma and I travelled together by plane from Glasgow to London and then onto Rio, Manaus, until eventually we were paddling up the creek towards the camp. She was dressed to kill, wearing flower-patterned leggings, designer T-shirt, huge floppy denim hat and dark sunglasses. When she smiled, her two new 'grown-ups' front teeth made her look like Janet Street Porter on a jungle tour! I couldn't help but laugh when she burst out, 'Look, Daddy, there's a monkey,' and she pointed at Pete who was screaming 'welcome back' calls to me from a tree overhead.

As we pulled into the bank Pete leapt into my arms, chuckled loudly, and then peed down my shirt with excitement. 'Oh Petie,' Emma howled with laughter as I humped the luggage and a bag of Barbie dolls out of the canoe and up the slope to the house. On the doorsteps Pipoca, one of our female woolly monkeys, climbed into Emma's arms. Emma could hardly support Pipoca's 20-pound weight and flopped onto the bottom step. Then Pipoca promptly peed all down *her* shirt– and now we both howled. Pipoca 'keeowed' and her other relatives plummeted from the tree tops to come in for a closer look.

'Wow, it's big, Dad,' said Emma as she looked about the house.

'Ugh – what's this?' she asked peering into a glass tank as two black and red false coral snakes disentangled their shiny bodies.

'Snakes,' I said, 'and you'll be feeding them tomorrow – they eat fish.'

'No way, José – not snakes!' She was instantly fascinated by the various vivariums dotted about the place where giant spiders, venomous snakes and alien-looking insects lived. The threadbare rug under my work table caught her eye quickly, too. 'What's happened to this Dad?' she asked with her face screwed up into another 'ugh' expression.

'It's rot, sweetheart – everything eventually falls to bits out here. It's the humidity and damp,' I said.

She thought about this for a couple of seconds and then asked in a worried way, 'Will you fall to bits one day Daddy?'

We made a video of her visit so that she could show her friends at school what the rainforest was like. She wanted to present it – be in front of camera – so we spent most days investigating the trails, looking into termites nests, bromeliads, spider's holes and one memorable afternoon she even climbed the biggest tower. Harnessed and roped, desperate to start the climb, I was a bag of nerves. I didn't want her to see my concern.

'Are you sure – absolutely sure – you want to do this?' I pleaded.

'Look, Petie and Yetta are already on top, let's get going,' was her cool reply.

It took ten minutes and she hardly stopped, yet it seemed like hours as I inched up the same corner post alongside her. At the top before I could tell her to wait, my little eight-year-old hauled herself over the edge and onto the metal deck.

'Wow – look at the view,' she exclaimed, not even breathless. Yetta immediately grabbed her legs, pinning her to the spot and Emma began to laugh hysterically – which set me off as well.

'Get the video out Dad – I want proof that I did it,' she said. I wondered what her mother would say when she watched it back home. With Emma sitting in my lap, our legs dangling over the edge of the platform, we talked about the different types of trees and took in the panorama of the rainforest. I was so proud of her and happier at that moment than I could remember being for years.

The day Emma was leaving I woke up to find her in my hammock with me. We cuddled for a long while knowing that soon we would have to load the boat and head back to Manaus and the airport. Outside the house her eyes filled with tears when, with Pipoca in her arms, she was saying goodbye to her. I too tried to hold back tears.

Our brief time together had been a painful reminder of our time together as a family and, although gone forever, I missed it terribly.

As we paddled down the creek, Emma waved a final goodbye to Antonieta, Steve and the animal gang. A small fish jumped into the back of the boat and we struggled to catch the tiny flapping thing to put it back into the creek. I told Emma that it was a small matrincha. 'Do you eat them? What's your favourite fish?' she asked, looking seriously at me for an answer.

'Well I like lots of different kinds, but I suppose if I have to pick one it must be tucanare. What's yours?'

'Oh easy peasy, it's cod in batter with chips and tomato ketchup!'

Back to our normal routine again! We made two more trips back to the Sataré area. The first was on 26 September, 1995, when we persuaded the chief to give us two of their marmoset pets to hand-rear back at my camp. They were a male and female, almost adult, and I hoped that we could breed from them. None of these marmosets had ever been filmed before and if we could somehow manage to record the birth of babies it would be a remarkable sequence in the film.

I built a special set complete with hollow tree inside and introduced the pair in October. The set had to be redressed every few days, but one thing we were not short of was foliage. The tiny monkeys loved it. No longer tied to pieces of string, as they were by the Indians, they ran, jumped and played to their hearts' content. On Christmas Eve, Antonieta and Stephen called me from the house to their enclosure. There was no doubt that the female was pregnant, her tummy was fat and her nipples swollen. The only difficulty was that we had no idea when she would deliver.

I knew that the gestation time was about 160 days but as I had not seen any courtship behaviour it was a guessing game. Discovering her pregnancy lifted our Christmas spirits. We were to pass the holiday at the camp and had decorated the house with a few streamers and a scraggy imitation tree. Festooned in tinsel, baubles, flashing lights and a fairy on top, it gave the house a festive feel. I sprayed aerosol snow onto the glass fronts of the snake vivariums and we had snowmen candles.

Christmas morning dawned under a blazing blue sky and the sound of toucans calling. By 8.30 am it was over 30°C. We made great ceremony of opening our presents. Mine to Stephen was a tin of low-fat spam lite, while he gave me a CD storage case and, inside, a strawberry-flavoured condom. Stephen had also prepared Christmas presents for all the animals and went out into the forest and gave each of them a boiled egg – their favourite treat. I then opened my present to myself.

It was a satellite telephone which I had bought when I was last in Britain. My near fall from the tower and Nick having almost trodden on a *fer de lance* snake persuaded me that it would be a good investment. It also meant that I could call home (on rare occasions at £3 a minute!). The weather conditions sometimes made satellite connection a bit hit and miss, but generally it was clear.

I tried it out on Emma, whose first words were 'Happy Christmas Daddy – how's Pipoca?' She was already using the waterproof compact camera I had sent her to take pictures and promised to send some out to me. Listening to her rattle away thousands of miles away made me feel dreadfully homesick and after we had finished talking I went for a walk in the forest with Pete to brood. It was Stephen's first Christmas away from home and he was determined to celebrate a true Scottish Christmas and Hogmanay. At one minute to midnight, he disappeared into the dark night. The Brazilian radio station rang the sound of Ano Novo and then Stephen knocked on the front door.

'Come on in – you've missed the bells,' I yelled.

'No, no, you have to come and let me in,' he replied.

I opened the door, he stuck out his hand to shake mine and then handed me a bottle of whisky and a lump of coal!

'I bet you a fiver that we are the only idiots in the Amazon forest doing this,' I said, opening the whisky. He put a tape of Scottish music on and Jimmy Stewart belted out *Scotland the Brave*, followed by Paul McCartney and the pipe bands on the *Mull of Kintyre*. The music echoed across the forest around Alligator Creek again and again until four o'clock.

The year ahead, 1996, was to be a busy one for me on the film front. Marmosets – working title 'Gremlins' – was coming along well. We had another expedition planned to the Sataré, and in May I would begin filming a new Granada Television geography series for Channel 4 about Amazonas.

In February my eldest daughter Charlotte came out to stay for a month. Charlotte is 23 years old and lives in London where she had recently received her BA in Modern European Studies. Charlotte's mother and I never married and, to cut a long story short, my lifestyle hadn't given us many opportunities to meet up and so we were both looking forward to getting together.

Charlotte's arrival on 1 February couldn't have been better timed for the expected birth of the marmosets. On 5 February I decided I had to start the long vigil of night-time birth-watch. The female marmoset's stomach was so big she looked as though she were going to explode. I had installed special fibre optic lighting in the tree-trunk nest hole and set the camera up at five o'clock in the afternoon. It poked through a cut-out panel in the back of the tree-trunk. Darkroom material shrouded the camera and lens to help to camouflage it, but the two marmosets seemed oblivious to its presence.

Every night I would look through the lens and monitor her movements. The nights were long and uncomfortable because a colony of ants had emerged in the floor under my feet and I constantly had to wipe away the biting beasts. With yawns and scratches I would try to convince myself that something was happening. By 14 February, Valentine's day, I was beginning to wonder if indeed she was pregnant or just obese. I even considered giving it a break for one night's decent sleep but decided against that – sod's law being what it is. At five o'clock I took my place behind the camera. As usual the marmosets entered their night-time sleep hole just before six o'clock. I watched as they settled into a nook facing the camera. The female couldn't settle – or was I imagining this? She couldn't, or wouldn't, lie flat and remained sitting bolt upright. Behind me, 150 feet away, I could see the candle lights flickering on the dining table in the house where Stephen, Antonieta and Charlie were eating. I was tempted to join them but decided against it. Twenty minutes later Antonieta arrived and suggested I go and eat in the house while she kept watch. I made her promise to flash a signal with torch light if she noticed anything unusual.

I settled myself in a chair by the window with a plate of chicken casserole and rice and chatted to Charlie. She has an undeniable talent for languages and speaks Spanish fluently. Within her first week she had mastered Portuguese pretty well and could converse with Antonieta easily. She was just beginning to tell me about her plans when I saw the flashlight blinking.

I bolted out of the house and up the hill to the enclosure and Antonieta said that the female had pulled strange faces a couple of times. Looking through the camera, nothing appeared different – and then she suddenly clearly grimaced with pain. She was in labour and this was the night!

The contractions came every four minutes for the first half hour as her face contorted and body convulsed with each one. It was astounding to be standing only inches from her and I couldn't help compare the scene to a human mother giving birth. However, unlike my frantic jitterings at Emma's birth, the male marmoset was fast asleep, only opening his eyes when the female moved against him in pain and then closing them again.

After 35 minutes, the contractions came every minute and she gripped the inside of the tree hole for support. The pain of the spasms was now making her whimper and I was desperate for her to get the delivery over quickly – facing camera please. After 54 minutes, a tiny head appeared, then two matchstick arms and the female began to lick the baby clean and help to pull her out and up onto her breast to suckle.

This tot was no bigger than my thumb – I was mesmerised by the scene. It made a peeping sound and now the male woke up, suddenly showing intense interest in what was going on. He sniffed the infant and I am convinced that his hands were trembling because of excitement. The female contracted again and gripped the tree – another tiny head emerged and ten minutes later she had the twins feeding on her milk. Twins are normal for marmosets, whereas most other monkeys give birth to single young. No wonder she had been in pain, the combined weight of the infants was about 20 per cent of the female's body weight. To put that into perspective it is like a human mother giving birth to a 25-pound baby!

The male's role was only just starting. Three minutes after the second twin appeared, the female contracted again and the afterbirth appeared. It was a grisly scene as the male tucked in and devoured it, his white face matted with blood. Then he turned his attention to the umbilical cords which he chewed off. The female looked exhausted and slept. I knew that, from the next morning onwards, the only time she would have anything to do with the infants would be to feed them. At all other times the male would care for them and carry them about.

Charlie came up to look through the camera and see the infants shortly after they were born. She was as thrilled as I was and after packing up all the equipment and leaving the monkeys to sleep we went back to the house and celebrated with two bottles of beer.

Chapter Eighteen

The People Who Were Fish

By April 1996, the marmoset film was really taking shape. I considered that we already had the star sequence – the birth of twins – in the can. But I still wanted to feature the smallest monkey in the world, the pygmy marmoset. The Granada Television geography series for Channel 4, to be produced by my partner Nick Peake, was a double blessing for once. Not only was I going to see parts of Amazonas to date unknown to me but we were going to spend at least two weeks filming a group of Indians as far west as one could travel in Brazil. The forest where they lived was also home to the pygmy marmoset.

We arrived at the Alto Solimões hotel after dark and began shifting 15 cases of equipment into the first-floor entrance. At street level there was a shop called Big Trousers (in Portuguese), which sold bicycles. The hotel owner, a young Arabic-looking man, checked us in. Nazio Yousef Waveldow was not a Brazilian, he announced. 'I'm white!' he added as an exclamation. The rooms were only a little more than white-washed cement cells, but at least they had old air-conditioning units that worked.

It was 18 May and we had just landed at the Colombian-Brazilian border airport at Tabatinga. We would be living with the Tikuna Indians whose women and children commonly go about with pygmy marmosets in their hair, where they eat head lice. I was reunited with Gordon, who was the sound recordist. Lisa Perrin, a bright and attractive 26-year-old, was to be the researcher.

The following morning we met Nino, a representative of the Maguda tribe. Maguda means 'People who were fish'. It was the first whites to encounter the Maguda 200 years ago who had given them the name Tikuna, 'Men with fierce faces'.

Nino certainly wasn't 'fierce-faced'. He was tubby, short, friendly and had a face that wore a permanent smile. Because of my years in Amazonas I have, sadly, come to distrust some first signs of helpfulness, but in Nino's case I couldn't have been more wrong. He had arranged everything for us and at midday a boat would be waiting to take us 75 miles south along the Amazon river, called the Solimões west of Manaus, to the

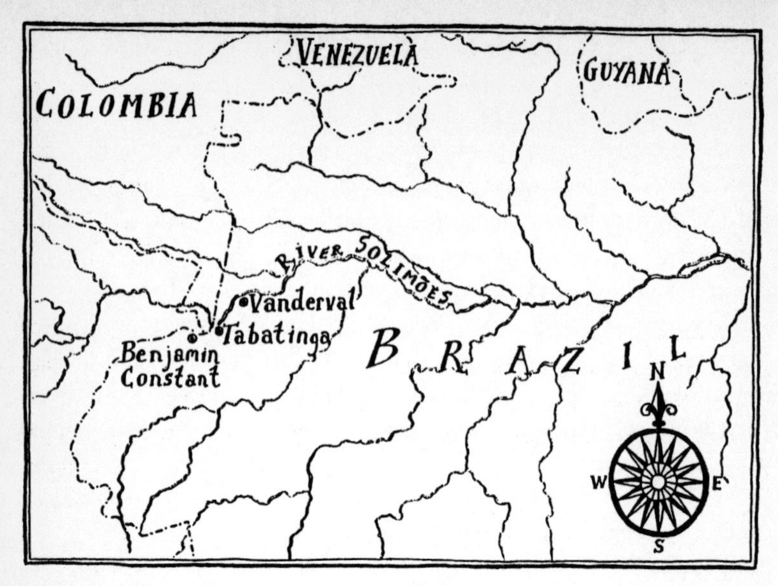

capital village of the Tikuna people. There we would meet their chief, Pedro Inacio. The driver was another Tikuna man, Tertulinho, who stood five-feet-nothing.

He would stay the two to three weeks with us and bring us back to Tabatinga. As we were leaving the hotel, a slick, slimy-looking young man approached me. He tried to rent me a light aircraft at $50 an hour, '*talvez gratis*' he threw in, 'maybe free'. We did want to do some aerial filming, but anyone willing to hire a plane out for $300 less than the going rate had to be suspect, especially on the border with Colombia.

Tabatinga port was a floating stage with a hut on it. The pontoon was crammed with small boats for hire. Hundreds of people bustled to and fro loading the boats, while street vendors plastered the rough earth roads. In the appalling heat, we hurried from store to store buying basic provisions for our journey, then carried them to the pontoon where Gordon and Lisa loaded them into the fast launch. While we waited for Tertulinho to load 130 gallons of petrol a young boy came over to me and pointed to my T-shirt. I was wearing a Boddington's Brewery design with the slogan 'By 'eck you look gorgeous tonight petal', and a Boddington's beer can. The boy told Gordon and me that his hobby was collecting beer tins, and that he had 147 different kinds. Sadly I couldn't add to his collection.

As we loaded the boxes of food and cases of equipment, it rapidly became clear that one boat would not suffice and we had to hire a second. It was a sellers' market and we were easily ripped off, but we had little choice. Tertulinho's boat had a 125hp engine on the back and it flew. We sped out from the port into the middle of the swollen river and went east, dodging flotsam at about 40 miles an hour. I closed my mind to the possibilities of collision. I thought, in any case, we were wearing life jackets and my cameras were insured.

An hour downriver we had to stop at a Federal Police post for them to search the boats and examine our permits. The area is a hot-bed of drug runners and no doubt we were prime suspects. I chatted with the officious looking men who looked like characters from an American cops programme. Their senior officer, wearing a black waistcoat and dark glasses, decided we were fair game. He told me that we had a '*problema sério*' after looking at our passports and visas.

'Yours is OK, but your friends' are not valid,' he informed me in Portuguese. He continued to explain that Nick, Gordon and Lisa could remain in the area and trade 'supermarket goods', but not have anything to do with filming. I might have laughed had I not understood Brazilian police methods. I realised, as he stroked his chin and looked at me, that he wanted money. I was determined not to allow him to get the better of us, but the matter was resolved by the soldier who had searched our boats. He told him we were clean and that 'it would be friendly to let them go'. After a minute of casual delay he did let us go, but said I had to take my comrades to the nearest consul and get their visas altered. '*Certo, certo sem dúvida.*' (Right, right – definitely!)

The largest Tikuna community is Vanderval where more than 1,000 Indians live. Its aspect from the river is pleasant, with thatched huts, mostly on short stilts, spread along the riverbank. As we arrived dozens of Tikunas clustered about us. Children darted forward to touch our arms before retreating into the throng, howling with laughter at having 'done the dare'.

Chief Pedro Inacio came down to the boats to meet us. This tiny man almost embraced us as we stepped ashore among the gathering crowds. He led us to his own house where he gave us two rooms to stow our gear and sling our hammocks. It was immediately obvious that a grand party was underway. Almost everyone had decorated faces, including young girls with painted cats' whiskers that gave them a cute but predatory look. A continuous sound of drum beats came from nearby. After we had organised ourselves the chief took us to the place where about two hundred Indians were gathered dancing. Four men were using carved sticks to beat two huge tracajá terrapin shells, which served as drums.

The molucca was large, about 70 feet long and 30 feet wide. Half way along one side was a thatched cell made from buriti palms and bamboo posts. Several men were busy decorating it with red and black dyes extracted from plants. A happy feeling permeated everything and although everyone stared, particularly at Lisa with her short, cropped blonde hair, they also smiled at us openly. We were offered a fermenting brew from a gourd, made from chewed manioc pulp. We didn't refuse, although I don't think Nick or Lisa realised at that moment that it was 'spit wine'. It was strong and Pedro seemed pleased that we had tasted it. The runs are on the way, I thought to myself.

I saw three young girls with pygmy marmosets on their heads and pointed them out to Nick. A group of women arrived escorting two completely cloaked figures. Pedro explained that these girls had just had their first menstruation and were being prepared for initiation into womanhood – the reason for the bash. They were taken into the palm cell which was sealed up behind them. They had to remain in there for 24 hours, without food or water, and could not sit or lie down. Three female elders remained with them to ensure the rules were obeyed. I can't imagine how they managed this!

As we left the molucca Pedro introduced us to some of the guests. A chief from a nearby settlement plonked a dead red howler monkey and agouti into Pedro's arms as a gift. 'Bit different to our bring-a-bottle parties babe, init?' Lisa commented.

An hour later Pedro had us all sitting on the floor of his house, eating with his family. One of his daughters, with a baby on her breast, gnawed on an agouti leg bone and a small child banged another one on the floor until its mother grabbed it from her and began to chew it herself. After dinner, we moved into the room we would sleep in, which was also their kitchen. In the smoky atmosphere the bushy grey tail from a dead saki monkey dangled over my hammock and the dead howler monkey lay stiff on the floor by Gordon's hammock, its eyes opaque. A mangy dog, snoring loudly, was curled up asleep under Nick's hammock. Bats zoomed in and out, grabbing insects. We had been told by Tertulinho that there were no mosquitoes in Vanderval; he had lied.

The noise from the drums went on all through the night and I couldn't sleep. Every so often an Indian or two peered through the broken doorway and giggled. At three in the morning I was overcome with severe gut pains and had to dash outside. I was being followed by three young men and decided to go into one of the pit latrines close to the chief's house. The hole was crawling with maggots and the smell so foul that I almost retched. I felt weak. Afterwards I went to the river in the dark, still followed by a group of young men, to clean myself up. When I returned to the room Nick was awake, Lisa was feeling ill and Gordon was snoring soundly. At 4.45 am his travel alarm clock went off. Nick shouted 'turn the bloody thing off' but Gordon couldn't hear that either.

At 5.45 am Pedro's son-in-law, Hildo, poked his head in the doorway and told us that the initiation rites would begin earlier than planned because the men were running out of drink! We were extremely weary. Looking through the open door of our room I watched a dog outside chewing on a monkey skull. We dressed and plodded with our gear over to the molucca to film the event.

There were hundreds of men, women and children, as well as dogs, parrots, macaws and pygmy marmosets. The majority were dressed in tattered rags and most of the children looked ill. There were many cases of conjunctivitis and I noticed that some of the smallest children had their eyelids stuck together with pus. The two young girls shut away the previous night were pulled out of the palm cell one at a time. They held their hands over their faces as though ashamed of their nakedness. A feathered head-dress was placed on both of them and then they were smeared all over with some plant extract that made their skin turn whiter. After this they disappeared once more into the cell. By the time they reappeared two hours later many of the men were paralytically drunk and falling about. Clearly the spit wine hadn't run out, or someone had produced more.

An ancient woman looked out of a nearby hut window and a few feet away from her a mother kicked her young child under the chin to stop it crying – it did. Maybe two years old, she had bloodied ear lobes from recent piercing and looked sick. A group of men brought a dustbin-sized metal pot almost full of the frothy hooch and the crowds gathered around to get their share. Pedro ordered two men to be physically dragged away as they seemed to be trying to kill each other. They were so drunk they simply kept falling into bystanders, including the camera and us.

The two girls eventually reappeared and were brought into the middle of the molucca, where they knelt on a palm mat. Immediately eight or more women gathered around them in a tight circle and began to rip the hair off their heads with their hands. It was a furious activity. The women doing it and the crowds watching looked as though they were enjoying it all, but the two girls sobbed loudly and tears of pain ran down their faces. Every so often an older woman would thrust a gourd of alcohol in front of them and they would gulp it down to deaden the experience. It took almost half an hour to pull every hair from their heads, then they were lifted onto palm stretchers and carried through the village to the beat of terrapin drums and singing from the crowds. We were told that the hair-pulling was the girls' penitence for disobeying their parents during childhood.

By lunch time my stomach was back to normal. Lisa made some reassuring corned beef and onion sandwiches and we ate them listening to the sound of animal fat spitting as something was being cooked on an open fire in the next room. 'I hope that's not the howler monkey for our dinner tonight!' I said to the others and we started to fantasise about what we would like to be eating. While they, more used to civilisation than me, talked of curries and lasagnes, I found inner happiness dreaming of thick-topped seafood pizzas and Arctic Roll pudding.

We returned to the molucca for the afternoon to continue filming. By this time our main entertainment was betting which of the Tikuna men would be able to stand up again after falling over. At one point three men appeared at the entrance of a large hut about 100 yards away and the crowds surged towards them. The men were passing out wooden sticks spiked through charred leathery looking fish. After half an hour hundreds of people were dancing around us waving these Amazonian kippers in the air. This ritual comes from their belief that the Tikuna were once fish that jumped out of the Solimões river and were transformed into people.

Later, down by the river, guests started to leave the village and paddle home. Only the women were capable of driving, while the men were laid out in the bottom of their dugouts, some snoring, some singing. Several of them staggered over to us to say goodbye, before falling into their canoes – at least I think that's what they were saying. One man, his eyes like dishes of fruit cocktail mixed with raspberry jelly, shook my hand five times before falling backwards onto the muddy shoreline.

Back in our hut a dog stumbled in. Someone had painted it with dye and had plastered feathers all over it and the poor creature rubbed itself against our hammocks to rid itself of the gum matted in its fur. Later that afternoon Nick and I found a ruined school house, presumably from when the missionaries were in Vanderval between 1972 and 1979. We played cribbage for an hour until Gordon and Lisa arrived with the ingredients for our dinner – tuna salad with cream crackers. At dusk the mosquitoes became so bad we had to return to our smoky kitchen back in the village. By nine o'clock we were in our hammocks desperately trying to find sleep. I lay there reciting a litany of complaints: 'I want a shower, I want to clean myself, why don't those children stop crying, why do the bloody cockerels think dawn comes every 15 minutes.' The sound of the Indians urinating in the small hours outside our hut turned my stomach.

I opened my eyes at 5.30 am. There, in the dim light, was a young boy, about five, sitting on the wooden floor by Nick's hammock chewing on the remains of our previous night's food. At 7 am mists shrouded the village and as we walked towards the river Pedro joined me and told me it signalled the time for the great river to begin falling. Gordon appeared, but as the chief dived into the water we decided against taking a plunge, having noticed some lumps of human excrement floating close to the shore.

Back in our hut I cleaned my teeth with a mug of mineral water. Hildo's wife was squatting by the fire cooking tambaqui fish. I looked closely at her. Her face was smeared with mud and her hands black with carbon, but she was extremely attractive. She looked over at me and smiled, showing off her teeth. They were all chiselled to points, her gums mostly black. Later that morning I spoke to Hildo for some time. Hildo was different from all the others. It was noticeable in his manner, his speech and his dress. At 28 years old, he spent most of his time in the town of Benjamin Constant on the other side of the river from Tabatinga. He had been educated by the missionaries and eventually trained in the town as a dentist. He explained that his wife and three children never left Vanderval. I asked him about the chiselled teeth and he explained it was a fashion among the Tikuna women. He laughed and went on in good Portuguese '*Eu vou contar um segredo para você*' (I will tell you a secret) and almost had me in a state of collapse when he told me his wife's chiselled teeth were prosthetics he had made.

While Hildo and I were chatting, a young girl spread strips of palm leaf on the earth to dry. Later the women would weave hammocks from them. Hildo told me that his father had died from being bitten by a *fer de lance* snake in the forest farm, less than five minutes after the bite. I told him about my expeditions to the Sataré and I explained about the tocandeiro ants at initiation ceremonies – he told me that Tikuna men also have a use for them, but for a slightly different reason. After a man and woman have come together, he said, sometimes the woman 'doesn't think the man's penis is tasty'. He explained further that this was not a reference to oral sex, which was, according to him, purely a white man's custom, but that perhaps the wife didn't think her man very good at 'it'. If this happened, the man had to put one of these ants on his penis and let it sting it. I felt my scrotum tighten at the mere thought.

Nick decided Vanderval was too big for us to film and wasn't in any case representative of how most of the Tikuna people lived. Chief Pedro arranged to travel with us into the interior to choose a community that would suit us better. At 11 am on 22 May, we turned off the powerful and brimming Solimões into a small river. We were heading for a Tikuna village called Tabatinga ('White Earth'), like the border city where we had flown into. We stopped first at another Tikuna settlement, Barro Vermelho ('Red Earth'). Pedro knew everyone and they greeted us, without exception, in surprisingly warm manner. The chief of Barro Vermelho was the shortest Tikuna man I had met, about 4 feet 8 inches. Apart from his heavily lined face, he looked like a little boy, wore wellington boots and grinned all the time. Baskets of abiu appeared for us to eat. These mango-sized yellow fruits were delicious, but after one or two too sweet for me. You bite off one end of the thick but soft skin and suck the pulp out with the seeds. These you spit out afterwards.

The small group of thatched huts that made up Tabatinga looked perfect for filming. We noticed that the further we were away from the main river and Vanderval, the more shy the people were. More baskets of abiu and inga fruit were brought for us to eat and while Pedro explained to the village elders what we wanted to do I explored the settlement with Nick. There were 26 thatched huts, all built high off the floor on stilts. At one end of the village near the river was the 'festa' house, a large molucca. It looked as though it had only recently been constructed or renewed, with its palm thatch clean and bright and the support poles clear of spiders' webs and ants' nests. A heavily pregnant girl passed us, a pair of pygmy marmosets on her head. She walked with a child to a hut where a razor-billed curassow pecked at the earth and a channel-billed toucan perched colourfully on the corner of the front steps.

We reached an agreement to film at Tabatinga, but were asked not to start work until two days later, so unfortunately it was back with Pedro to Vanderval and an invitation to dine with him. Nick and Lisa rather freely said they couldn't face it, so to avoid any disrespect Gordon and I took up the invitation. Hildo's wife was sitting on the floor with everyone else noisily sucking the brains out of a monkey skull; we were late for dinner it seemed. We sat next to Pedro and ate just a little of the meat, while he told us some of the Maguda's history. He had us spellbound as he recounted how the Indians were massacred by the conquistadors but recovered their numbers over the following 100 years. Then the white man arrived and massacred them again, but this time the Indians captured one of the attackers, cut off his head, boiled him and ate him. After that, Pedro said, there was no more trouble. This was the incident that led to these normally friendly people being given the new name of Tikuna ('men with fierce faces'). Before the massacres there were 300,000 Tikuna; now there are only 25,000, yet they are still the largest single group of indigenous people in Brazil today.

After dinner Nick and I played cribbage in the village food store. Shortly after we started dealing the cards 21 tiny Tikuna children crowded about us, captivated by what we were up to. They kept touching our feet and legs and mumbling to each other in hushed voices. Hildo appeared and explained that the kids were fascinated by the softness of our skins – theirs tended to be toughened by long exposure to the sun. The room was reduced to hysterical laughter when a giant bufo toad hopped in front of our cribbage board, disrupting our game.

That night I slept well until four in the morning when familiar gripping stomach pains struck. At 4.15 it seemed that every dog in Vanderval began to bark, then the roosters joined in. I heard Lisa scream from the next hut 'shuuuuuut uuuuuup', but I guessed they only understood Tikuna. At six o'clock, despite us still lying in our hammocks, Hildo's wife and kids came in and lit the fire. Smoke billowed up into the dark brown stained roof. They sat on the floor and scooped burgundy-coloured açaí palm juice from gourds into their mouths. It dripped everywhere. Eventually the smoke grew so thick it became almost impossible for us to breathe, so we packed our equipment and rucksacks and waited by the river for our departure to the small village of Tabatinga.

By the time we arrived there I had severe stomach pains and Hildo disappeared into the forest behind the village to find a '*curo*' for me. The way I was feeling I would have taken anything to get rid of the pains so his brew concocted from a tree's bark was simple to swallow. Half an hour later my entire intestinal system had had the equivalent of a 'T' Cut job, and the pains didn't return. Gordon made a corned beef salad for lunch while we spread out our equipment in a spacious hut made from pachuba palms.

That first afternoon we filmed an interview with Pedro. One question Nick put to him about whether his people could adapt to the changing world about them brought an answer that said it all for me!

'We have gone through our biggest changes already in the past 200 years,' said Chief Pedro Inacio Pinheiro. 'First when the conquistadors arrived and massacred our people and then when white man arrived and more or less did the same. But we recovered, except white man's religion has killed a lot of our own culture.'

The mosquitoes were terrible in the hut and I slung my hammock under the molucca tree to escape them. Fifteen Tikuna men squatted in a group just 15 feet away from me and stared at me without talking for almost an hour. It began to unsettle me a little. By nine o'clock they had gone and I lay there listening to the sounds of the forest, looking up at a brilliant starry sky. Three dogs ambled into the shelter, but because of my white mosquito net they couldn't see me. I switched on my torch to stop one of them peeing on the bottom of my net and gave all three of them a hell of a fright. They ran as fast as they could to a nearby hut and barked furiously in my direction. Tikuna voices shouted and cursed their racket until they finally shut up.

At 5.15 am the cockerels started crowing and in the first flicker of daylight I saw a man and a young girl of about 12 sharpening their machetes on stones. Then they walked towards the river, the man also carrying a fishing spear.

At six o'clock it was fully light but the village was cloaked in a heavy mist. Somewhere nearby a parrot argued with a rooster and an old man came into the molucca and stood over my hammock staring into it for a few minutes. Another man with badly scarred legs and only one eye went past carrying a shotgun, while two scraggy dogs had a tug of war over the skin of a dead saki monkey. I went over to the hut where Nick, Lisa and Gordon were sleeping to find them still in their hammocks, but awake and cursing the night's mosquitoes. Lisa's legs looked as though she had measles.

We needed to film the way the Tikuna farm, fish and generally survive in the forest. The second day we went with one family to their farm. Their plantation bordered pristine forest and we walked among manioc, banana and pineapple plants. Three small children, aged two, three and five, all wielded machetes and did their bit. It was amusing watching the youngest boy. He pulled and pulled to try and get a manioc tuber out of the hard ground, then the root would suddenly give and he'd fall onto his bottom. Then he would get up, pick up the razor-sharp machete and clear the ground around the next plant. We snacked on an inspiring lunch of tinned sardines and dry biscuits and went down river to film a man hunting fish with a spear. Despite having watched locals hunt this way for several years I still cannot see the fish they launch their spears at. On one occasion a man kept saying '*olha olha*,' look look, there it is – but all I could see was muddy water.

Back in the village towards the end of the day, two young girls, about three and four, and a tiny boy who couldn't have been more than two, were paddling enormous dugout canoes. The little boy was on his own in a canoe more than 12 feet long! Their small naked brown bodies twisted and turned as the girls raced the boy up and down the river. As they struggled to beat each other, a small hawk briefly appeared out of the forest with a snake in its talons. It saw the humans and darted back into the trees. The sky turned aquamarine blue and the clouds were edged in bright orange which darkened as each minute went by.

On that third night, as I settled my hammock in the molucca, Hildo's voice came out of the darkness, 'Who's sleeping in the house of parties?' he enquired in Portuguese.

'It's me, Nick' I replied. 'You're brave,' he added and walked away. I began to wonder why he should think I was brave. The dogs were barking sporadically and I knew that jaguars were known to attack them from time to time, but surely not here, I told myself. A group of five young lads were walking nearby when my ropes slipped and deposited me on the floor. I heard them laughing their heads off.

On our fourth day we planned to film one of the Indians hunting in the forest. We pulled the boat into the entrance of a narrow creek where there were two thatched huts. One of them was a food store and the other the living area. A hammock was swinging gently to and fro with a bundle in it – a tiny new-born baby. This was where our hunter lived, but sadly the sequence got off on the wrong foot. We asked the Tikuna man to walk past the camera on his way into the forest and then to return so that we could film it from another angle. It's ridiculous, but it is the way with film-making. Either his Portuguese or his hearing couldn't have been that good because he never returned! We didn't see him for another four hours.

We had much better luck with two young men and their women when we followed them into the forest where they intended to fish – but not with hooks or spears. They were first going to find and dig up the roots of the *shimbo* liana. After half an hour the Tikuna stopped by the base of an enormous fallen tree. Its root base towered over our heads at 15 feet and I paced out the width, 30 feet across! The four Tikuna began to scrabble about in the earth nearby and slowly began to expose a network of roots. They cut these into foot-long lengths and piled them into carrying baskets. They worked away gathering the roots for an hour and I estimated that they must have pulled at least 150 feet of the root free in that time. What didn't fit into the baskets was bundled together with strips of tree bark. We then set off deeper into the forest as they carried the roots on their heads.

We eventually came to the edge of a shallow clear-water creek. The men immediately began to clear a small area of vegetation. A woman cut two ten-foot-long poles, about five inches in diameter. These poles were placed on the leaf litter and used as anvils. They fashioned four hammers from other wood and spent the next hour smashing the roots they had gathered into a pulp. An unpleasant smell quickly permeated the area and as I filmed the close-ups, Hildo came to my side and told me to make sure I kept my mouth shut. The splattering of the fibres over the camera and into my face hadn't bothered me until he pointed out that just a few specks in my mouth could kill – it is that deadly, he assured me. Later he told me that some Tikuna have committed suicide by putting the root pulp into their mouths. I had never considered that these people might suffer such mental anxiety or depression that they would want to end their lives. I thought suicide was our 'civilised' world's property.

Thunder began to roll in the distance. One man looked up into the trees and then moved away from the others. He returned a few minutes later with an armful of large wild bananeira leaves and in less than ten minutes had constructed two small shelters that they call '*rabo do jacu*' or tail of the guan. A short while later, while the Tikuna continued to prepare the root pulp, it started to rain. It was not just slight rain, but a 20-minute deluge and it made me appreciate how skilled the Tikuna are at detecting the forest's moods and dangers.

Hildo shouted across to the other Tikuna and I asked him what he had said. 'I told them I was staying here with the white man in his hotel.' We all laughed. When the rain had stopped one of the two women appeared from beneath her shelter with a newly-woven basket of palm leaf spine. This was to carry the fish home in. Looking at the apparently lifeless creek water, that was optimism, I thought.

When the root had been pounded into mulch, it was separated into brick-sized bundles and wrapped in palm leaves. One of the men had dammed the creek nearby using palm leaves and wooden poles. Suddenly we were on the move again and the going became much harder. We stumbled along the creek edge and veered away into dense undergrowth. After 15 minutes we came to the same creek again, only much farther upstream. The two young men strode into the knee-deep water and started to swirl the parcels of poisonous mushed roots about. White clouds of sediment rose to the surface and drifted extremely slowly away with the current. We made our way back to the dam wall with Hildo while the two other men followed the creek.

When we arrived nothing appeared different: certainly the white poisonous clouds hadn't discoloured the water here yet. However, within just a few minutes, fish began darting past us as they came downstream to escape the venomous sediments. When the fish reached the palm wall they turned and went back towards an inevitable death. It was amazing how quickly they were stupefied by the poison. As the colour of the creek began to cloud a little, these fish immediately floated to the surface and were carried by the current to the dam. Here one of the women gathered them up, threading them through their gills with a thin liana before placing them into the basket. Considering the small size of the creek there were some surprisingly large fish. We were thrilled to have seen this happen and although uncomfortable and exhausted we set off on the return journey happy in the knowledge that we had great images in the can.

On the way back down river I noticed smoke filtering through the canopy on the far bank and then saw a group of Tikuna in the edge of the forest. About 20 men were in the process of making a 40-foot long dugout canoe. The tree trunk had been hollowed out and, propped up along its entire length with sticks, turned over so that the cavity was sitting over a fire of the same length. We spoke through Hildo and asked their permission to film.

The fire under the trunk had to be carefully controlled, the heat softening and swelling the wood to allow it to be shaped properly. Too hot and the wood would burn, too cold and it wouldn't shape. They all pulled, turned and grunted together as the trunk was then rolled over to expose the hollow interior. The cleft through which they had carved it out was extremely narrow, about eight inches wide, and the men began to fit pre-prepared wooden forks along both edges of the slit. Liana ropes were then tied from these back to staves in the floor. Halfway along the lianas were posts that could be turned like a windlass to stretch and tighten the bush ropes and so open up the trunk wider.

All the activities we had been so privileged to watch that day left us greatly impressed by the Tikuna's ingenious labours. These skills, handed down from one generation to the next, illustrated just how resourceful these people still were.

We stopped at the hunter's house again and all had a laugh as Hildo explained to him what had happened. It seemed the hunt had gone well for him and a very large, male, grey saki monkey hung from his shoulders, the monkey's tail tied to fashion a carrying sling.

He invited us to sit and eat with his family and we squatted on the earth outside the farinha house. His daughter came and poured boiling water over the monkey and pulled out all its glorious thick fur. She then plonked it into a large cooking pot, its hands and feet drooping over the edge. It looked much like a little naked human baby, but in the blink of an eye it was cut up and over the open fire. A young man brought over a large metal bowl, half full of açaí juice. Despite the fact that I really don't like the stuff I scooped a little into a gourd just to be polite. I find that the liquid dries out my mouth and leaves an unpleasant aftertaste. We were all given a piece of charred fish to eat with farinha. I could almost feel the usual stomach ache coming when the daughter appeared and handed Tertulinho a cooked arm from the saki monkey – he was to take it back to Chief Pedro as a present. On our way back it was bizarre to see this little arm and perfect hand sticking out from between a bundle of green leaves in the bow of the boat.

On our last afternoon at the small village we went down to the river to bathe. It was so much nicer here than at Vanderval simply because there were fewer people with their inevitable pollution. Two young boys were in the middle of the river paddling a large dugout canoe. They waved and smiled at me and I waved back. Then I noticed some fallen abiu fruit on the ground and decided to throw one at them. It landed at the edge of the canoe, sending a plume of water up next to them. They immediately grabbed the floating fruit and threw it back. I stupidly tried to catch it and the abiu exploded in my hand, spattering my head in the sticky sweet pulp. War was declared and Gordon, Nick, Lisa and I lobbed every fruit we could gather. Hildo was there and shouted, 'Indians against Whites,' while a young Tikuna boy started to help us to collect the fruits. Villagers began to gather to watch the foreigners' antics. Hildo shouted again, 'White man has put a worm into our boy's head!' and the crowd roared with laughter. Some of our missiles found their target but most didn't. The boys shrieked with laughter and once we had thrown all of our ammunition at them they retaliated. Considering their age, size and that they were standing in an unstable canoe, they were excellent shots. It was good fun and afterwards we all swam together in the river as the sun went down.

The following morning I woke up at six and looking through the white mosquito netting I could see a small boy standing under the molucca. I lifted the shroud up and smiled at him. He looked absolutely shocked and pointed his finger at me and then screamed at the top of his voice while running away. I knew that I hadn't shaved for a few days, but surely, I thought, I couldn't have looked so awful that morning.

We loaded the boat with our equipment and set off back to Vanderval. Chief Pedro was handed the monkey's arm and said we should eat with him this last night. Nick and Lisa looked at me and I smiled. They couldn't get out of this one. As usual we all sat on the wooden floor. Two oil lamps burned, giving smelly illumination. The chief and his family ate with their fingers, while we used spoons. We tucked into smoked fish that tasted like varnished sandpaper, rice and farinha. I covered mine with our own pepper sauce to mask the taste. The company, however, was wonderful.

Nick, bravely I thought, gave a speech to the chief in Portuguese and I thanked him for accepting us into his family. The chief suddenly started singing. He told us when he had finished that it was an old Tikuna legend.

'Hundreds of years ago when the White conquerors came to Magudaland they murdered many, but many soldiers were killed, too – about 5,000. The soldiers also killed almost all the forest animals for food. Those animals that survived went into a cave underground at the base of a mountain. While some of the animals, like the jaguars, kill people, they allowed the Maguda survivors to enter the cave for protection. The soldiers gathered thousands of pepper plants and made them into a powder, which they put on the ground around the hole and set fire to it. The animals and people couldn't get out and the fumes burnt their eyes badly. Eventually the soldiers went away because they couldn't find any more animals for food and the Maguda and the animals of the forest emerged and prospered again.'

The chief told us this is why the men were beating the terrapin shell drums at the party, as remembrance and gratitude to the animals that helped them to survive. It was a perfect last evening with the Maguda and it left us all with warm memories of our time there. We had been impressed by their friendliness, hospitality, resourcefulness and simplicity.

Leaving Vanderval wasn't quite as straightforward as we may have hoped. For some reason there was a bit of a bash in the village that night and we only managed a little sleep amid the revellers' din and the clouds of mosquitoes.

By eight o'clock, we were standing by the riverside, the sun already burning fiercely in the clear blue sky. The usual crowd of onlookers were there while we waited and waited for Tertulinho to appear. We hadn't reckoned on him being one of the party animals, but when I found him in one of the huts he was still drunk. Anxious to get back to Tabatinga and a hotel where we could wash and relax before our evening flight back to Manaus, I gave him two cans of Coca Cola to help him to sober up. Then the engine wouldn't start and it seemed that everyone knew what was wrong and tried to put it right.

The chief finally appeared, staggering under the weight of a 15-gallon fuel container. He was in an awful state, too. He grabbed a bottle of *cachaça* from someone and took a big slug from it, then sat down in the shade of a hut, put his head in his hands and went to sleep. Eventually, an hour and a half later, the engine was fixed and we waved goodbye as we motored out into the fast-flowing Solimões river.

Chapter Nineteen

Giant Snakes and Birds That Go Moo!

On 9 July, I was greeted back at Jacaré camp by Petie, who jumped into my arms, gripped me and giggled, then peed down me as usual. Having finished filming for Granada's geography series, my next task was to complete the final seven months of filming marmosets.

Almir, Antonieta's brother who had replaced our *caseiro* (handyman) Claudio, arrived and, pointing to a cage at the water's edge, said that he had a surprise for me. He began to tell me that only last night Antonieta, Stephen and himself had been sitting in the house eating dinner at seven o'clock when a terrible scream echoed through the camp. Almir ran down to the peccary enclosure on the edge of the creek to find our male, Poomba, being constricted by the powerful body of a giant anaconda. He shouted to Stephen to fetch my revolver and, unfortunately, rather than firing the gun into the air to frighten it away, he shot the creature mid-body three times. Amazingly, the snake survived and despite the flesh wounds seemed to be in good shape.

Poomba was none the worse for wear, although he was reluctant to go near the water for a while, like the rest of us who lived at the Jacaré. We couldn't believe that we had been swimming there for three years with that monster lurking nearby! Almir managed to get a rope around the anaconda's head and, together with Antonieta and Stephen, dragged it out of the peccary enclosure and put it in a large portable transport crate that we have. I could hardly believe its size. Its body was as thick as my thigh and from nose to tail it was 12 feet long!

We talked about building a water pit for it so that we could keep it a while and possibly film it, but I knew that I really wanted to let it go as soon as possible. As we walked up to the house I noticed that Almir had some fresh scarring on his chest. While I had been away with the Tikunas he had been trying to fix a distant neighbour's home-made shot gun. A cartridge had exploded and he had been lucky not to have suffered worse consequences. When the young man who had built the contraption returned later in the day he was horrified to see the bloody wounds on Almir. For a joke Almir told him that he had had a fight with my assistant Stephen and in a fit of anger had shot him dead. The boy ran to his canoe and paddled off quickly. The following morning he was back and Almir asked him why he had run away. '*Eu tive medo, mais agora eu voltei para ajuda a enterrar*! I was scared but I have returned to help you bury the body,' he said.

Early the next morning I was filming a fishing hawk eating a fish on a branch over the creek when my vision went milky and I almost fainted with the most severe abdominal cramps. That night I got into my hammock after starting a new course of amoeba medicine but the pains were still excruciating. Five times I had to get to the bathroom, tripping up the stepped but steep muddy hill behind the house. I just wanted to be in a clean normal bed with a tiled bathroom and a hot shower. Deciding to stay in my hammock and rest for a day, I went to my bookshelf and pulled out a Robert Goddard novel to find that potter wasps had taken it over. My copy of Kipling's *Jungle Book* had suffered a worse fate, termites had eaten their way through the first three chapters. I finally settled down with a musty-smelling Isabel Allende novel, *Eva Luna*, an absorbing read that took my mind off my ills.

By the end of July I felt well again and filming the marmosets was back on schedule. A friend of mine, Ron Kiley, a police inspector from Glasgow, had offered to bring Emma over to my camp for her summer holiday rainforest visit. Ron was the constable on the island of Mull and we shared a common passion for photography and watching wildlife. They arrived just in time to witness the birth of our second marmoset twins, which Emma thought the highlight of her stay.

While Emma and Ron were with me the BBC Natural History Unit in Bristol asked me to do a reconnaissance in a forest reserve operated by the Smithsonian Institute 70 miles north of Manaus. One of the Unit's producers, Peter Bassett, arrived to join me on the expedition to look for a calf bird lek (courtship) site and it was agreed that Emma and Ron could accompany us.

Courting calf birds have never been filmed and they are truly extraordinary creatures. Many times in the forest behind my own camp I had heard their strange calls. The first time I thought there was a young cow on the loose in the forest because they sound just like a calf calling for its mother, but at other times like a mewing cat. At their lek sites the males gather and display to attract a female by calling and puffing themselves up and sticking out their rumps. It is a most bizarre sight and sound, one of nature's great shows.

We arrived at the Smithsonian's camp late in the afternoon after a bone-shaking 25-mile drive through the forest on a bush road. A Brazilian ornithologist who was familiar with the site accompanied us into the forest to show us roughly where the birds lekked. The site was a hard one and a half-mile trek from the camp and I realised that Emma's little legs wouldn't stand the pace. It was rough terrain, up and down severe hills, so I asked Ron to take her back to the camp and await our return.

'I'll come with you if you carry me, Dad!' she pleaded.

Peter, Sergio, the ornithologist, and I took an hour and 20 minutes to hike to the area. We heard one distant bird which excited us, but we knew that the main time for lek activity was at dawn. We agreed to get up at five o'clock and see what was happening there at sunrise. We set off at a fast pace in the half light, tripping over tree roots and fallen branches. The air in the forest was moist and the damp smell of rotting vegetation hung everywhere. The only sounds apart from our feet squelching or snapping wood were stridulating crickets and peeping tree frogs, but soon the birds began to wake up. A few bats whizzed overhead on the way to their daytime roosts and a mot mot bird started calling, 'hudu hudu'.

After the first mile into the forest we heard the distant but unmistakable mooing of three calf birds. By the time we reached the area where they were calling I was knackered, and Peter, wiping the sweat from his face, equally shot. Sergio loped along with sound recorder and binoculars around his neck as though he were out on a gentle summer stroll, but his ears pricked at every bird sound. It has always amazed me how people with this talent can pick a slender-billed tyrannulet out from a flavescent warbler, but they can.

Peter and I squatted on a convenient log while Sergio prepared his playback equipment. Sweat dripped from my forehead into my eyes and stung. I looked at Peter and his glasses were all steamed up. Then a bird close by cruelly called out 'free-beer, free-beer, free-beer'. 'Fly catcher – yellow crowned tryannulet,' Sergio said, while watching it with his smart new Baush and Lomb binoculars.

Play-back is just that: a previous recording of a bird calling is played back to try and bring the same species of bird in close. More often than not it works, but what wasn't working at this moment was Sergio's equipment. He was trying to fix the badly rusted terminals of his batteries to a small amplifier. Eventually after scraping away some crud, I had to stand there with my thumbs keeping the contacts together. Then the recorder produced the clear sound of a single calf bird calling. Almost instantly a bird replied and we saw the shape of it fleetingly in the top of a tree overhead. A minute later another called, and then another! Within ten minutes we had 12 calf birds displaying 40 feet above our heads. The cacophony was remarkable, their strange courting cries penetrating far into the dense forest around us.

The expedition to the Smithsonian reserve was a total success and we left having arranged for me to return in late November to film a sequence of the calf birds at their lek when the males would be in full flight trying to attract a female. The sequence, if I could get it, was to be included in the new David Attenborough series, *The Life of Birds*. Both Peter Bassett and I were extremely excited at the prospect of getting what could be one of the 'nuggets' of the series. Only time would tell.

On the morning of 6 November, as we were going down to the creek at Jacaré to wash, I heard a bleating call coming from the deer's enclosure in front of the house. To our great surprise we found a tiny fawn curled up under a palm frond near the gate. Pete and one of the woolly monkeys were investigating this new family member by pulling her ears. A well-aimed stick sent them scurrying up into the trees and Flower, the fawn's mother, came to its side and licked her. She was only a few hours old but managed to stagger to her spindly feet to suckle. We had no idea that Flower was pregnant and had simply put her large frame down to being overfed.

It was a week of unexpected arrivals. Five days later one of our adult female woolly monkeys, Lucy, appeared on the steps of the house with a tiny furry bundle clamped onto her chest. Her baby looked like a puppet ET, its wrinkled face giving it the appearance of an 80-year-old. She wobbled her head trying to focus as I stroked her and was probably wondering who on earth this thing was that mum was letting touch her. Lucy had given birth during the night and a long piece of the umbilical cord still dangled from the baby. This string of drying flesh attracted Pete who naturally decided it was there to be pulled. As all the other monkeys gathered around Lucy to inspect the new arrival, some of them simply sniffed while others gently touched tentatively with their fingers. Pete, however, became even more excited when he realised that each time he gave the cord a yank it produced a squealing noise! Lucy put up with the first two pulls and then let Pete know who was boss. He galumphed across the open ground in front of the house chuckling loudly and, climbing a tall tree, rested while he waited for Lucy to go to sleep knowing that then the game could start again.

The rains came early this year, a blessing after the trials of the last dry season when it would take a back-breaking six hours of hauling the canoes over dried-out river beds to reach the camp from the main river. On I November, the first of the season's storms woke all of us in the house just after three in the morning. We fumbled for torches and tied down the tarpaulins to keep out most of the driving rain. Cracks of thunder directly overhead made us jump with fright, and the monkeys peered in at us as they huddled against the window grills. In the midst of the tempest we heard a tremendous crashing noise that sounded as though one of my scaffold filming towers had fallen over. It could only have been the tower that I had built in the middle of the creek in front of the house, the others were too far away to be heard, and I was worried that it might have fallen onto Gordi the tapir's enclosure. I foolishly dashed outside to investigate and slipped down the muddy slope to where I could hear him whistling. The slope was awash with rivers of rain cascading towards the creek and I fell sideways into a jauari palm tree.

Twenty-seven of its needle-sharp three-inch spines buried themselves in my left arm and hand. It was agony. I immediately pulled out as many as I could but the pain was excruciating. The tower hadn't fallen over, it was the scaffold store behind the house that had collapsed under the weight of the deluge and Gordi was galloping about in the rain having a ball.

Back in the house Stephen and Antonieta set to digging out the broken-off points of the spines with tweezers and razor blades. Two of them, however, had gone deep into the muscle of my forearm and broken off. By the following morning I had lost the use of the fingers on my left hand and any movement caused agony. Two days later I was in a hospital in the city having my arm cut open by a doctor who obviously hadn't heard of keyhole surgery. After an hour of him prodding the slit that he had cut, he announced that he couldn't find even one, and suggested that I have some x-rays. He stitched me up while seven other local patients, waiting their turn for treatment, stood over me gabbling to each other about how dangerous the 'poisonous' jauari spines were. I began to feel a little faint until the doctor turned to one of them and said that he didn't think this palm was venomous.

The x-rays didn't show up anything, but he gave me a course of pain killers and antibiotics and suggested I would be better waiting until I got back to Britain because his radiography machine was too old – 'over 30 years' – he seemed to say with some pride. 'I live here,' I said, but was already thinking it would be a great excuse to go home for Christmas.

It had been five years since I had met Rick West, although we kept in touch with each other regularly by mail. He had wanted to visit my camp for a long time – the area was an unknown where tarantulas were concerned – but his work and family commitments had conspired to prevent a trip until now. I received a message saying he could get to Amazonas for five days and could I meet him at Manaus airport on 9 November? The plane arrived, but Rick didn't and to my utter horror a message was waiting for me that he was in detention at Sao Paulo airport in a federal police cell!

I called and was put through to Rick. He was devastated. 'They're deporting me tonight, guy, because I don't have a visa.' Rick explained that the agency which had arranged his travel had told him he didn't need a visa to visit Brazil, and he had asked them on two separate occasions just to be sure. I phoned his wife Lynn who, despite being extremely upset for Rick, coolly said that she was on her way to the travel agent. 'Don't worry Nick, he'll be with you next week.' And he was.

The travel company had compensated Rick for its mistake and seven days later I was holding a flashlight in the forest while Rick rooted under fallen logs for tarantulas and bugs. Five years had done nothing to dull his sense of humour and we spent a tremendous five days together discovering spiders' haunts that I had walked over without seeing for the past three years. It was as though it had only been a few weeks since we had been together and the two highlights for both of us that week were the discovery that

Theraphosa blondi lived in the forest next to the camp and that, in a log only ten feet in front of the house, he even managed to find a new species of tarantula.

'Jeez, guy, you mean you have seen this almost every night for years?' he asked incredulously when I told him we all knew it lived there.

One evening, before his regular nocturnal bug hunt, he looked at the photos of my family and friends on the hut wall and asked what the immediate future had in store for me. I explained my ten long years of yearning to make a film about jaguars and the other smaller but little-known cats of Amazonia. It seemed somehow fitting that, with the filming of the marmosets coming to an end, the crossroads I knew I was approaching should be discussed with Rick. He had been there almost at the start of the long-term Amazonian experience and, while in Venezuela together, had helped to fire my ambitions to make films in the great forest. Despite knowing my passion for making wildlife films he detected something else within me, which he tactfully pointed out.

I was staring at the photo (encapsulated in plastic to prevent mould) of my daughters, Emma and Charlie, that had been taken in a London restaurant the year before. Their faces beamed out, brimming with happiness, and Emma had her arm around my shoulders.

'You know, guy, you can never get those moments back,' Rick said philosophically. 'As much as I love the rainforest – and this place you have is a paradise – I just couldn't survive the loneliness here and I don't know how the hell you can. I think you're ready to hang your jungle boots up.'

'Jaguar first, Rick, if Survival commission it, and then home for a spell – perhaps the life of a sequence cameraman?' I replied, trying to imagine living at home again and being torn between two worlds.

I knew that in my heart I had made the decision and only wondered if I would ever be able to readjust to life back in England or Scotland. That discovery, however, would be two years away. First, I wanted to find and film the elusive and most powerful animal in the rainforests of the Amazon – the feared and revered jaguar was my next ambition.

Index